A GUIDE
TO METAL
AND PLASTIC
FINISHING

M. L. M A R O N E Y

A Guide
to Metal
and
Plastic
Finishing

I N D U S T R I A L
P R E S S   I N C

Library of Congress Cataloging-in-Publication Data

Maroney, M. L.
A guide to metal and plastic finishing / M. L. Maroney.
174 p. 15.6 × 23.5 cm.
Includes index.
ISBN 0-8311-3032-6 :
1. Metals–Finishing 2. Plastics–Finishing. I. Title
TS653.M36   1991
671.7—dc20   91-6278
                                                        CIP

Industrial Press Inc.
200 Madison Avenue
New York, New York 10016

First Edition

A GUIDE TO METAL AND PLASTIC FINISHING

Printed and bound by Thomson-Shore, Inc., Dexter, MI

2   4   6   8   9   7   5   3

First Printing

Preface

A Guide to Metal and Plastic Finishing is a compilation of general information relating to the finishing of metals and plastics: information concerning buffs, abrasives, abrasive belts and wheels, and polishing and buffing compounds. In addition, suggestions as to their usage and their methods of usage on metals and plastics are presented.

This information, both general and specific, is for the entrepreneur and the professional, and for use in the job shop and the in-plant metal and plastic finishing department.

Frequent suggestions are made to consult with your supplier when in doubt. If you do not have or know a supplier, refer to the yellow pages of your local telephone directory or to the Thomas Register. Suppliers may be listed under "Buffing Equipment and Supplies," "Metal Polishing Equipment and Supplies," or "Plating Equipment and Supplies." If technical information on metals is needed, we refer you to *Metals Handbook – Desk Edition*, ASM International, Metals Park, OH 44073.

The methods and processes discussed herein are for general information only. The author assumes no liability for any misuse of this information.

A c k n o w l e d g m e n t s

Acknowledgments

I wish to sincerely thank the following companies for their help and suggestions in the writing of this book.

Acme Manufacturing Co., Madison Heights, MI 48071

American Iron & Steel Institute, Washington, DC 20036

Artistic Brass, Inc., South Gate, CA 90280

AutoSpin, Carson, CA 90748

B & H Industries, Mason MI 48854

Bacon Felt Co., East Taunton, MA 02718

Barker Brothers, Inc., Ridgewood, NY 11385

Boyle-Snyder Plating Co., Hollywood, CA 90028

Brownells, Montezuma, IA 50171

Bob Burton Plastics, Hesperia, CA 92345

C & G manufacturing Co., Baldwin Park, CA 91706

Coast Metalcraft, Rancho Dominguez, CA 90016

Committee of Stainless Steel Producers, American Iron & Steel Institute, Washington DC 20036

Cosmo Whells Division, Chicago Rubber, Winchester, KY 40391

Cyro Industries, Woodcliff Lake, NJ 07675

Divine Brothers Co., Utica, NY 13503

Felton Brush Co., Manchester NH 03105

Gruber Systems, Valencia, CA 91355

Hermes Abrasives, Inc., Virginia Beach, VA 23452

Highland Plating Co., Hollywood, CA 90028

Hy-Tech Spinning, Inglewood, CA 90302

The Manderscheid Company, Chicago, IL 60606

Matzie Golf Co., El Segundo, CA 90285

Modern Plating Co., Los Angeles, CA 90045

The Norton Co., Worcester MA 01606

Reliable Buff Co., Monroe NC 28110

Rohm & Haas, Philadelphia, PA 19105

The Sattex Corporation, White City, OR 97503

Siefen Compounds, Inc., Wyandotte, MI 48192

The Spartan Felt Co., Spartanburg, SC 29304

Speed-D-Burr, Vernon, CA 90058

Sunnen Products, St. Louis, MO 63143

Swest, Dallas, TX 75220

United States Pumice Co., Chatsworth, CA 91311

WLS Coatings, Los Angeles, CA 90061

Western Abrasives, Vernon, CA 90058

Warnings

GRINDING—POLISHING—BUFFING—COLORING IS DANGEROUS!!!

Using a grinder or polishing/buffing lathe as small as a 1/4 hp can be and IS dangerous.

BE CAREFUL: KEEP YOUR EYE ON THE BALL!!!

BE SURE—that your machine has the proper hoods!

BE SURE—to wear eye protecton. Your eyes are your PRECIOUS possession!

BE SURE—to wear a dust mask or respirator!

BE SURE—that the SFM (surface feet per minute) is correct!

BE SURE—that the wheels are in balance! This is important!

BE SURE—that you work below the center of the wheel!

BE SURE—to start slow and careful! Speed improves with practice!

BE SURE—that you do NOT wear loose or baggy clothing!

BE SURE—to keep your attention on what you are doing!

BE CAREFUL!!!

Glossary

"Acme" or "Acme Finish"	Descriptive tradename of a lime compound used in buffing nonferrous metals. (No longer available.) Term used to describe a specific type of finish (now seldom heard or used).
buff	1. A wheel used to buff with; 2. the act of buffing; 3. as described in this book, the third operation in the finishing of metal or plastic.
buffer	1. One who does buffing; 2. describes the machine—handheld or up-right—used for handbuffing.
buffing	Act of doing or applying a specific type of finish to metal or plastic.
buffing brick	A mild abrasive and grease mixture in bar or brick form or liquid used for buffing.
"Butler" finish	Term applied to a type of finish on metal by an abrasive and glue mix material such as "Lea" or greaseless compound.
cement	A liquid adhesive used "cold" as opposed to a "hot" glue.
"cold" glue	Refers to cement.
electroplating	Electric deposition of metal (shorter term "plating" more commonly used).
emery	An abrasive grain, usually black in color; may refer to American or Turkish emery,

which are black; may refer to aluminum oxide or silicon carbide.

emery cake A mixed compound containing silica and emery used as a lubricant on polishing wheels.

emery, Turkish A natural abrasive material obtained from Turkey and ground into various mesh sizes.

ferrous Iron or steel, or an alloy with iron.

glue An adhesive made from animal hides and bones.

"hot" glue Glue that is melted and used heated.

grain Ground abrasive material.

grease In metal and plastic finishing, tallow or petrolatum that has been combined wih materials to make a hard material for use as a lubricant on abrasive belts, disks, or wheels.

grease, to Describes an operation or step at the beginning of finishing metal.

grease stick A mixture of greases in a paper tube.

greaseless Term or name applied to an abrasive and glue mixture used on a wheel for mild cutting of metal. Used to describe a "grain" or "satin" finish on metal.

greaseless compound Sometimes called "Lea" or "satin" compound. Occasionally, the name "Butler finish" is heard describing the same kind of compounds.

grinder Machine used for grinding or removing metal from metal; also used to describe person doing the grinding

grinding Used to describe the first step in metal finishing; method used to remove metal from metal.

grinding wheel The hard abrasive wheel made of vitreous or abrasive grain.

HP Horsepower; the power rating of any motor, electric or otherwise.

jeweler's rouge An iron oxide and grease combination in bar or liquid form and used for the final finishing

	operation on metals or plastics; synonymous with "red" rouge.
nonferrous	Referring to metals without any iron or steel (aluminum, brass, copper, etc.).
polish	Routinely refers to act of finishing or finish desired on metal or plastic.
polisher	Refers to person doing the finishing or polishing; also may be used to refer to person buffing.
polishing	The second operation in the finishing of metal or plastic; also general term used to describe finishing or shaping of metal or plastic.
polishing wheel	Canvas or cotton wheel on which an abrasive has been bound wih an adhesive
polishing wheel cement	See "cement"
rpm	Revolutions per minute; the number of times a wheel turns per minute.
rouge	Describes any buffing compound, bar or liquid, used for the final finish on metal or plastic; may be "red" or jeweler's rouge, chrome rouge, or sometimes "tripoli" rouge.
rubber	Erroneous term used to describe buffer or polisher.
rubbing brick	Refers to bar buffing compound.
rubbing compound	Sometimes used to describe bar buffing compound
sfm	Surface feet per minute; the number of feet covered per minute by each revolution of the outside or periphery of a wheel.
sanding	Term used to describe the using of "sandpaper" or "sanding belt" or "disk."
sandpaper	term used to describe the paper to which an abrasive has been bound by an adhesive.
"set-up"	The condition of a wheel to which an abrasive has been bound, i.e., "a set-up."
set-up wheel	Refers to the wheel itself used to "set-up" or bind an abrasive with an adhesive such as a cement or "hot" glue.

stainless bar   Referring to a type of compound to be used on steel, stainless steel, or sometimes aluminum.

tripoli   A dark brown or light tan bar or liquid buffing compound for use on nonferrous metals or plastics.

tumbling   A kind of operation or method used in finishing metal or plastic.

white diamond   A silica-based compound used in the finishing of metals or plastics.

white tripoli   Sometimes referring or meaning a silica-based compound, bar or liquid.

Metals

Metals	Pure Metal	Alloyed Metal
1. aluminum	aluminum	
2. beryllium	beryllium	
3. brass		brass
4. bronze		bronze
5. cadmium	cadmium	
6. chromium	chromium	
7. cobalt	cobalt	
8. copper	copper	
9. diecast		diecast–zinc or aluminum
10. gold	gold[a]	
11. iridium	iridium[a]	
12. iron	iron	
13. lead	lead	
14. magnesium	magnesium	
15. nickel	nickel	
16. nickel silver		nickel silver
17. osmium	osmium[a]	
18. palladium	palladium[a]	
19. pewter		pewter
20. platinum	platinum[a]	
21. ruthenium	ruthenium[a]	
22. rhodium	rhodium[a]	
23. silver	silver[a]	
24. stainless steel		stainless steel
25. steel	steel[b]	
26. tellurium	tellurium	
27. tin	tin	
28. titanium	titanium	
29. white metal		white metal
30. zinc	zinc	

[a]The eight precious metals.

[b]Steel is arbitrarily classified as a "pure" metal, because it is made from iron ore, and is not considered an alloy.

Note: While the pure metals may be used as such alone, or alloyed with other metals, they are all used for specific purposes. While there are other metals, these listed ones are those we are primarily concerned with either in their pure form or as an alloy in other metals. Refer to *Metals Handbook*, ASM International, Metals Park, OH 44073.

Contents

Preface	v
Acknowledgments	vii
Warnings	ix
Glossary	xi
Metals	xv

Section 1. General Information

1. Establishing the Shop—"Job" and "In-Plant"	3
2. Buffs	5
3. Wheels	16
4. Wire Brushes	19
5. The Work-Holding Spinner	21
6. Abrasives	23
7. Abrasive Belts, Cartridge Rolls, Disks, and Sandpaper	28
8. Contact Wheels	30
9. Use of "Hot" or Animal Glue	34
10. Polishing Wheel Cement	36
11. Greaseless Compounds	38
12. Kool-Kut	40
13. Buffing Compounds	42
14. Liquid Buffing Compounds	46
15. Cleaning	54
16. Buffing and Polishing Machines	59
17. Summary	69

Section 2. Metals

18. Aircraft—Private   73
19. Aircraft—Commercial   76
20. Aircraft Windshields   78
21. Aluminum—to be Dyed or Anodized   80
22. Aluminum—Sand Castings   82
23. Finishing Aluminum Sheet   84
24. Barrel, Tumbling and Finishing   86
25. Brass—Sand Castings   87
26. Brass—Sheet and Tubing   88
27. Bronze Castings   90
28. Burnishing   92
29. Cast Iron (Grey Iron Castings)   93
30. Deburring   95
31. Diecastings   97
32. Electroplating   99
33. Precious Metals   101
34. Gold   102
35. Sterling Silver   104
36. Silverplate   106
37. Guns   108
38. Honing   113
39. Lapping   114
40. Marine Propellers   115
41. Metal Spinning   117
42. Nickel Silver   121
43. Nonferrous Sand Castings   123
44. Stainless Steel   125
45. Vibratory Finishing   129

Section 3. Plastics

46. Plastics   133
47. High-Melting-Point Plastics   134
48. Low-Melting-Point Plastics   136
49. Acrylite—An Acrylic   139
50. Plexiglass—An Acrylic   141
51. Cultured Marble   143
52. Fibreglass   145

General Information

Establishing the Shop—"Job" and "In-plant"

The "job" shop, also known as the "polishing" shop, may be an independent shop or an important adjunct to an electroplating plant. Such shops do work for anyone—an individual, a retailer, an interior decorator, or a manufacturer. On the other hand, the in-plant shop is a part of or a department in a particular factory.

While the in-plant shop usually works on one type of metal, the job shop is normally equipped to work on any metal: ferrous (iron and steels) or nonferrous (aluminum, brass, bronze, copper, diecast, pewter, gold, and silver). (Some job shops specialize on working with precious metals only.)

Both the "job" and the in-plant shop must comply with all federal and state safety and environmental regulations. If one is planning on establishing such a shop, it is advisable to become familiar with all such regulations and restrictions. Also, when the necessary equipment has been purchased and installed, only a minimum of supplies should be purchased, to provide an opportunity to pretest wheels, buffs, compounds, abrasives wheels, or belts for quality of finish, ease of cleaning, and costs.

Once your equipment is ready, all buffs, wheels, and supplies should be marked and identified as to their specific use. If using "set-up" or polishing wheels, mark each wheel with its mesh or grain size and an arrow to indicate the direction the wheel is to run. Also, if glue is to be used, use a thermostatically heat-controlled electric glue pot as appropriate and a round glue brush for each size grain, and keep the brushes clean when not in use.

Store the grain and glue/cement containers so that they do not become contaminated. Keep the containers covered when they are not in use.

In addition to polishing wheels, abrasive belts can be used, since they can save time and speed production.

It will be necessary to have a dirt/dust collection system. Such a system consists of hoods or guards for each polishing lathe, the associated piping, and a fan to blow the dirt/dust into a suitable receptacle, such as a "cyclone," so-called because of the twisting path the dirty air takes through it. A competent supplier specializing in such blower or exhaust systems should be used to advise you of the volume of air needed to keep the shop clean and to engineer and install such equipment in order to comply with all governmental regulations.

Always use safety equipment—dust masks, respirators where necessary or advisable, guards or hoods on all equipment—and have adequate ventilation and lighting. Be neat—have a place for everything and keep everything in its place.

Buffs

Introduction

Buffs are wheels manufactured from disks (either whole or pieced) of bleached or unbleached cotton or woolen cloth used as the agent for carrying abrasive compound during buffing.

The quality of the cotton used in buffs affects the life and productivity of that buff. This quality varies, due primarily to the water and its mineral content used in the processing of the cotton. Thus, it is suggested that testing be performed before buying buffs in quantity. When buying in quantity, determine the composition of the cloth, whether it is 100% cotton or a mixture of cotton and polyester; testing which composition of cloth holds the buffing compound better and which provides the better wear.

Your supplier can recommend the proper buffs to use—these recommendations will be based on the shaft speed and the power of the buffing lathe, the kind of material to be buffed (metal or plastic), the kind of buffing compound to be used, and the finish desired.

There are two kinds of buff treatments: the treatment used by the manufacturer of the cloth (cotton or sisal) at the mill or a dip treatment used by the buff manufacturer. The objective of either treatment is to increase the life and productivity of the material.

Treating sisal buffs—the untreated sisal deteriorates over time—preserves the sisal and also improves the "cut." The untreated sisal buff softens under the frictional heat generated in use, more so than the treated sisal. Here, again, prior testing is rewarding.

Types of Buffs

Buffs (sometimes referred to as pads) may be made of cotton, cotton or wool flannel, canvas, chamois, sisal, or string. They may be loose, sewed, "airways," fluted, or pocketed, and made of plain (untreated) or treated materials. The following list briefly describes buffs:

1. Loose buffs—several plies of material having one row of sewing around the arbor or center hole.

2. Sewed buffs—spiral or concentric (single row of sewing) as close as 1/16 in. to as far apart as 1 in. or more.

3. Airway, or as is sometimes called ventilated—usually 16, 18, or 20 plies of material pleated or "puckered" by being wound around a drum and being constricted by a clinch ring center.

4. Fluted—wherein the cloth is "ruffled" instead of being "pleated" and where the cloth is folded to look like fingers. (This buff is not very well known.) (One type is manufactured by Barker Bros., Inc.)

5. Finger buff—strands of cloth wound in such a way as to form "fingers" held by the clinch ring in the center. It is a very aggressive buff.

6. Pocket buff—having disks of material folded to form "pockets" and having several rows of close sewing around the arbor and several rows a little farther out.

7. Sisal buff—a strong cloth made of sisal or hemp (for example, "burlap" or rope). This buff type may be plain or treated, full disk, or "ventilated" having the metal clinch ring in the center. It is a very aggressive buff used on ferrous metals or heavy nonferrous castings.

8. Chamois buff—a very soft material that is used primarily for buffing the precious metals. This buff may be loose or sewed.

9. String buff—made of cotton strings on a wood or metal hub. A very soft buff, superb for using with a satin or greaseless compound. It may be used to "wipe off" plastic or precious metal to avoid leaving buff or thread marks, also used in shoe factories or shoe repair shops to wipe off shoes.

In addition to the preceding buffs, there are jewelry buffs, razor- or knife-edge buffs, round- and square-end goblet buffs, taper buffs, cylinder buffs, and many special buffs, and solid felt wheels and "felt bobs."

Buffs are sold by the "each" or by the "section": an "each" may be a buff of any number of "plies" or "layers" of cloth, such as 20, 40, 60, 80, or 100 plies (100 plies is the maximum thickness the buff sewing machine can sew). Buffs may be ordered in "soft," "medium," or "hard" densities. The density specified usually refers to the thread count of the cloth. The thread count may be 60/60, 64/68, 80/92, or 86/93. These numbers refer to the number of threads running vertically and horizontally in the cloth:

the larger the thread count, the stronger and harder is the cloth. The material or cloth may be "bleached (white)," "unbleached," or "treated" (the treatment most often provided is a "yellow" treated material). Other treatments in various colors are used by buff manufacturers to identify the type and purpose of the specific treatment.

The following illustrations and descriptions are of the individual buffs and their uses.

Miniature buffs (Fig. 1) are made of 100% cotton, usually of 64/64 thread count; however, the 80/92 material also makes an excellent buff. The diameters may be as small as 1/2 in. to as large as 3 in., thicknesses vary from 16 to 80 plies. The arbor sizes range from "pinhole" (1/8 in.) to 3/8 in. The sewing may be in 1 row around the arbor or in rows 1/8 to 1/4 in. apart. These tiny buffs are used on the small electric or air-operated and handheld equipment, or on a very small special bench lathe.

Figure 1. Miniature buffs. *(Courtesy of Reliable Buff Co., Inc.)*

"Jewelers' buffs" (Fig. 2) may be loose buffs (having one row sewn around hole), in diameters ranging from 3 to as large as 10 in. (Larger diameters are available by special order.) These buffs usually are stocked in 6 and 8 in. diameters, 60 and 80 plies, loose and concentric sewed in rows 1 in. apart, 64/64 or 80/92 unbleached and yellow-treated muslin; they have a leather, hard shellacked, or lead center. Jewelry manufacturers and silver/gold plating shops are large users of this type of buff. The concentric sewing allows for cutting the sewing when the buff wears or a more flexible small buff is needed.

The razor- or knife-edge buff (Fig. 3) is available from 2 1/2 to 6 in. diameter, unbleached or yellow treated, with a leather center. (It is available

Figure 2. Jewelers' buffs. *(Courtesy of Reliable Buff Co., Inc.)*

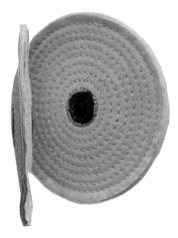

Figure 3. Knife-edge buff. *(Courtesy of Reliable Buff Co., Inc.)*

in 8 in. diameter on special order only.) These buffs are used extensively in the jewelry trade. The cloth is 64/64 in the unbleached, 86/82 in the yellow-treated variety. They are very handy to have in the shop.

Loose buffs (Fig. 4) are usually made of 64/64 unbleached muslin and are most commonly used as "coloring" or "finishing" buffs, which is

Figure 4. Loose buff. *(Courtesy of Barker Bros., Inc.)*

Figure 5. Sewed buff. *(Courtesy of Barker Bros., Inc.)*

the final buffing operation prior to lacquering, painting, or plating. The loose buff can also be used to buff soft plastics. Diameters may range from as small as 3 to as large as 16 or 18 in., in 18- or 20-ply thickness, with the arbor as desired or needed. These buffs are best when made of 100% cotton.

The sewed buff (Fig. 5) may be made of scrap or pieces of cotton, or of full disk 64/64 unbleached cotton. Sewing may vary from as tight as 1/8 in. to single rows 2 or 3 in. apart to a 1/2 in. spiral sew. These sections average 1/4 in. thick and may be "stacked" to form a wheel of any desired thickness. Sewed buffs may be used for buffing operations on nonferrous metals and plastics or as coloring buffs, but they are being replaced by airway buffs.

The pocket buff (Fig. 6) is made of full disk pieces folded to form "pockets"; it may be used or mounted one way for a "cut" with a tripoli or white diamond abrasive and mounted the other way for use as a coloring or finishing buff. Pocket buffs come in 64/64 unbleached or 86/82 material.

Figure 6. Pocket buff. *(Courtesy of Barker Bros., Inc.)*

The string buff (Fig. 7) is made of cotton strings on a metal or wood hub. This buff is an excellent "satin" wheel (using greaseless or "Lea" compound) for use as a "wiping" buff on plastic or precious metal. This buff is also excellent for coloring precious metals, since it leaves no buff marks. It is available in 4–16 in. diameters, face widths of 2–6 in., and arbors of 1/2–1 1/4 in.

Figure 7. String buff *(Courtesy of Felton Brush Inc.)*

"Airway," "ventilated," and "bias type" are the words used to describe this buff (Fig. 8), where the cloth is pleated and held by a metal clinch ring in the center. This buff is available in 4–24 in. diameters with centers of 3–9 in.; the cloth, for the most part, comes in 86/82 thread count, although it is available in 64/64 cloth. The amount of cloth in each section is designated by #2, #4, or #6: the least amount of cloth being in the #2 class with each succeeding class containing a larger amount of cloth—the added cloth making for a tighter and more aggressive wheel. Also, the buff is available in various treatments (see your supplier). Sections can be stacked to create a wheel with a "face" to meet a specific need.

Figure 8. Airway buffs. *(Courtesy of Barker Bros., Inc.)*

The sisal buff (Fig. 9), sometimes referred to as the "straw" buff, may be made of laminations or layers of alternating cloth and sisal, or of full disk material, plain or treated. "Airway" sisal buffs (Fig. 9b) are also available. In the full disk and laminated sections, the sewing is a spiral pattern 1/8 in. for a hard, stiff buff or as wide as 3/8 in. and, occasionally, 1/2 in. The sisal buff is very aggressive and primarily used on ferrous metal. The sisal buff is also frequently used on nonferrous metals, particularly sand castings. Before use, the sisal buff may appear to be too rigid and too hard; however, the buff softens and becomes more flexible under the heat generated during the buffing operation.

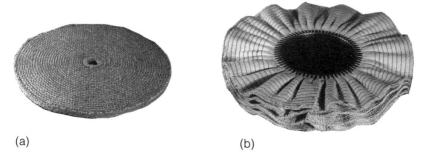

(a)   (b)

Figure 9. Sisal buff. *(Courtesy of Barker Bros., Inc.)*

Cloth polishing wheels (Fig. 10) are made of sections of sewed buffs stacked and glued or cemented together to the face width needed. The sections used may be of either full disk or pieced material, the sewing determining the hardness of the wheel. Sections made of full disk material almost guarantee a balanced wheel, while sections made of pieced materials have to have each section balanced against the next section to ensure a balanced wheel.

Figure 10. Cloth polishing wheels. *(Courtesy of Reliable Buff Co., Inc.)*

As noted previously, the sewing in the buff sections determines the hardness of the wheel: for instance, sewing 3/8 in. apart provides a medium soft wheel, sewing 1/4 in. apart provides a medium hard wheel, and 1/8 in. apart provides a very hard wheel. For the average shop, it is suggested that a mix of the three be stocked, unless the shop is working on a continuous flow of the same item requiring a specific type of wheel.

Other polishing wheels are made of compressed canvas and wool. The canvas wheel, and to a lesser degree the wool wheel, is very expensive. The canvas and wool or wool felt wheels are manufactured in several degrees of hardness: soft, medium hard, hard, and rock hard.

Your supplier should be consulted when selecting polishing wheels.

The goblet buff (Fig. 11, middle) is made for buffing the inside of bowls, mugs, cups, etc., or any inside surface needing both the sides and bottom finished. These buffs are made of cotton muslin and may be set-up with abrasive grain or greaseless compound, with a cutting compound, or with a rouge. The sizes manufactured are 2, 2 1/2, 3, 3 1/2, 4, 5, and 6 in. diameters, with a pinhole center, a round end, or a square end. They are used on taper spindles.

Figure 11. (Left) Square-end buff; (middle) goblet buff; (right) taper buff. *(Courtesy of Barker Bros., Inc.)*

The cylinder buff (Fig. 12) is used for buffing the inside of tubes or hollow ware where it is not necessary to polish the bottom. These buffs are made of cotton muslin and are available only in 2, 2 1/2, 3, and 3 1/2 in. diameters, with a hard pinhole for use on taper spindles.

Figure 12. Cylinder buff. *(Courtesy of Barker Bros., Inc.)*

The taper buff (Fig. 11, right) is similar to the cylinder buff except for its shape. These buffs are made of cotton muslin and are available in 2, 2 1/2, 3, and 3 1/2 in. diameters, with a hard pinhole center for use on taper spindles.

The cylinder and taper buffs may be set-up with abrasive grain or greasless compound or used with tripoli, a rouge, or other buffing compounds.

The razor edge buffs (Fig. 13) are a very handy item to have on hand: they are useful for getting in between items such as the tines of a fork or certain jewelry items. These buffs are available in 2 1/2, 3, 4, 5, and 6 in. diameters in a good grade of unbleached cotton muslin or in the harder yellow-treated cloth, with a reinforcing leather center for use on the taper spindles.

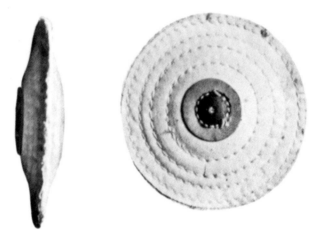

Figure 13. Razor-edge buffs. *(Courtesy of Barker Bros., Inc.)*

Bobs and "Points"

There are many instances where there is a need to polish the inside of a die or mold; to polish an intricate, complex casting; or to polish items or parts that cannot be polished on a regular polishing lathe. Some of these operations, using hand-held electrical or air-operated tools or flexible shaft machines, may require the removal of stock or imperfections, burrs, and so on. For such operations, we have what is called "points": bonded abrasives in the various shapes as illustrated in Fig. 14. These points are available in aluminum oxide and silicon carbide and in grits from coarse to very fine; they are mandrel mounted, in diameters as small as 1/16 to as large as 1 1/2 in. (Check with your abrasive supplier.)

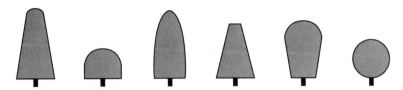

Figure 14. Points and bobs.

Following a polishing with a hard bonded abrasive point, the next need might be to buff to a high finish; for this need, there are the felt "bobs," available in the same shapes as the bonded abrasive points. The bobs are made of 100% wool and come in medium, hard, and very hard densities and mandrel mounted, in essentially the same sizes as points. The bobs may be set-up with glue and abrasives, and they may be used with greaseless or "satin" finish or with various buffing compounds. The designations for the felt bobs are standard plain or standard shaped felt bobs. (Check with your supplier.)

On occasion, a small wheel is needed to polish or buff in difficult to reach areas or where working space is at a premium: for this situation, "square edge" wheels are available. These wheels are made of 100% wool and are available from 1/2 to 5 in. diameter and a 1/8 or 1/4 to 1 in. face; arbors are available from a pinhole (1/8) to 1/2 in. and in five hardnesses. These wheels may be set-up with glue and abrasive or greaseless compound or used with buffing compounds. (Check with your supplier.)

Wheels

Introduction

The following information is general in nature and is offered to the grinding, polishing, buffing, or color wheel user.

As with any wheel or piece of equipment, be sure that the necessary flanges and guards are in place. When starting a wheel, allow it to run for a few seconds before using it, thereby allowing any loose material in the wheel to be thrown off. The following paragraphs give brief descriptions of the various wheels and their uses.

Grinding wheels are composed of bonded materials, such as aluminum oxide, silicon carbide, or diamond. They are very hard and may be referred to as "stone" wheels. Since these wheels generally operate at a high rpm, they can break, or "explode," so it is very important that guards or "hoods" are in place, that the wheel has the proper flanges when mounted, and that the user or operator wears goggles. These wheels should be operated as recommended by the manufacturer.

Polishing or "set-up" wheels operate at lower speeds than the grinding wheels. These wheels may be made of cotton, canvas, wool, or leather; the most frequently used wheels are made of sections of sewed cotton buffs glued or cemented and pressed together to a required dimension. Here again, the use of flanges and the placement of guards or hoods are necessary to a safe operation.

Tampico wheels are made of a natural fiber, and are often used in the untreated form with an abrasive such as emery cake or a buffing compound for light deburring or finishing. When properly treated, the wheel may be used for heavier and tougher jobs. These wheels hold compounds well and, while not as well known today as in the past, they still have their special uses.

Wire wheels, made of steel, stainless steel, brass, or nickel silver wire, have many uses: removal of rust or scale, burrs, or paints and lacquers.

They may also be used for a scratch brush or "satin" finish and for "high-lighting" an oxidized finish on brass, copper, or silver. (Safety goggles must be worn to protect the eyes from loose wires.)

Buffing wheels are made of sections of loose cotton buffs, sewed cotton buffs, or unbleached cotton muslin; they may be untreated or treated and airway or "ventilated" buffs. These wheels may be one section or several sections held together on the shaft by steel flanges. The buffing wheel should be allowed to run free for a few seconds before using. These wheels operate at a speed of 5000 to as much as 8000 sfm, so it is necessary to have hoods in place. These wheels can also "grab" parts, so be sure that they are used properly.

A relatively new wheel being used today is the "Scotch-Brite" wheel manufactured by Minnesota Mining: this name is used to identify a wheel, disk, and pad made of a nylon mesh material into which an abrasive—aluminum oxide or silicon carbide—has been blown. These wheels are used for light deburring, light polishing, or "satin" finishing. (Note, since these wheels are rather expensive, it is important that they are used properly.)

It is suggested that all wheels be marked with an arrow showing the direction they are to run. Using an old admonition, do NOT go against the grain!

Operating Procedures

First, a warning: a grinding, polishing, or any buffing wheel can be DANGEROUS when in use. GO SLOW! WATCH WHAT YOU ARE DOING—KEEP YOUR EYE ON THE PART! Go slow—speed increases with practice.

The wheel should turn toward the operator. While some wheels have an arrow on the side showing the direction they are to run, others are not marked. To determine the direction the wheel should run, lightly rub your finger along the face or edge of the wheel—backward and forward. Note that in one direction the surface is smooth, while in the other direction the surface is or appears to be rough. Do this several times to be sure which is which: then mark the wheel to run in the smooth direction. For safety reasons, also check the marked wheels in the same manner. (Another way to describe this is called the "flow" or "lean" away from you—as the "flow" in your hair: combing the hair from the forehead toward the back is the "flow," combing from the back of the head is against the "flow.")

If the wheel is mounted and run in the wrong direction, you will know it: the wear of the wheel is rapid, and it may even disintegrate! A grinding wheel may cause the part to chatter or bounce; a polishing wheel may knock the part out of your hand or cause the set-up to break away from the wheel; a buffing wheel will grab or knock the part out of your hand and also cause the cloth to fray.

Hold the part you are working on in such a manner that in the event the wheel grabs or knocks the part from your hand, the part will be thrown down and away from you—it is far better to lose a part than to incur an injury! (See Fig. 15.) To ensure that if a part is grabbed or knocked out of your hand and thrown away from you, always work below the center, as shown in Fig. 15a.

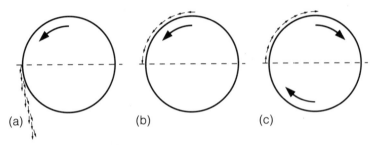

(a)   (b)   (c)

Figure 15.   (a) Using the wheel below center (arrows indicate the path a "grabbed" part would follow), which in the event of a grab, throws the part down and away from you; (b) correct "flow"; (c) incorrect "flow". (The operator is standing to the left.) For safety reasons, always be sure that the "hood" or "guard" is in place when using the wheel.

If necessary to change wheels for another operation, be sure that the wheels you are removing are marked with an arrow showing the direction they were and are to be running.

For safety reasons, always be sure that the "hood" or "guard" is in place when using the wheel.

Wire Brushes

We start this chapter with a safety admonition:

ATTENTION—WEAR SAFETY GLASSES!

When using mechanical brushes of any kind—wire, tampico, horse-hair—always protect your most precious possession: your eyes.

This chapter briefly reviews the use of wire brushes: brushes for cleaning, for deburring, for providing a specific finish on metal such as a satin or "brushed" finish, or for highlighting an oxidized finish.

The jewelry industry uses brushes of stainless-steel wire for mild deburring, for cleaning small and intricately shaped parts, or for highlighting on oxidized parts. Such brushes, approximately 1/4 or 1/3 in. thick, may range from as small as 1 or 2 in. in diameter to as large as 6 in. diameter, using wire 0.005 in. thick. In addition, the jewelry industry uses brushes made of brass or nickel silver wire, crimped or straight, with 0.005 in. thick wire.

The job shop will use a variety of brushes: steel wire, crimped or straight, 0.010 or 0.014 in. thick, for cleaning rust and dirt from parts; these brushes may range from a 6 in. diameter to as large as 16 in. diameter, and from 1/3 to 2 in. thick or wide. In addition, such brushes may be used to clean sand castings or as a first step before further polishing. Another use for such brushes is to clean threads.

Other brushes commonly used are brass and nickel silver wire, crimped or straight. These brushes may range from as small as 1 in. diameter to an average 6 in. diameter with one row of wire up to a maximum of six rows, with average wire thickness of 0.005 in. Such brushes may be used for wet scratch brushing of the gray scum common to silver plating, to apply a scratch brush finish, or to highlight an oxidized finish.

Other brushes available are knotted wire wheels, cup brushes, and end brushes—the knotted wire brushes being used for very rough and dirty work, the cup brushes being used mostly for interior work of various sorts, with the end brushes being used to clean and deburr borings, etc.

Whatever the need may be for brushes, contact your supplier for assistance.

Note: When using any type or kind of brush, ALWAYS follow safety and operating instructions of the manufacturer, particularly the proper sfm (surface feet per minute) or the recommended rpm (revolutions per minute).

The Work-Holding Spinner

The work-holding spinner (Fig. 16) can be one of the handiest tools to have in the shop. (Note the block of wood to which the flange on the end of the spinner is attached.) Do not confuse this tool with a spinner from a spinning lathe, which makes, for example, goblets, lamp shades, bowls, etc.

Figure 16. A work-holding spinner.

The spinner is used during the polishing and buffing of such items as metal lamp shades, bowls, conical or cylindrical objects, or round flat pieces of metals. These are items that have been spun and that can be held on blocks of wood formed to their shape; a flange, fastened to the wood, is screwed onto the end of the spinner. Then, the item is held against the buffing wheel to be polished or buffed; the speed at which the part spins is regulated by the pressure of the thumb. Buffing in this manner results in a uniform finish. This applies whether the part is having a spinning lines removed by an abrasive belt or wheel or whether a "satin" or "grained" finish or a highly lustrous buffed finish is being done.

The work-holding spinner is available in three lengths: 5, 7 1/2, or 9 in.: the handiest size being the 7 1/2 in. These spinners should be available from your supplier; it is manufactured by The Manderscheid Co., Chicago, IL 60606.

Abrasives

Types of Abrasives

Abrasives are substances that are used for abrading applications as in grinding and polishing. Abrasives are classified as grains and flours or powders. Some abrasive grain materials are:

aluminum oxide

silicon carbide

emery—American and Turkish

garnet

boron carbide

diamond—natural and synthetic

Some abrasive flour or powder materials are:

alumina

aluminum oxide

silicon carbide

iron oxides

silica

diamond—natural and synthetic

tripoli

tin oxide

zinc oxide

pumice

ruby dust

boron carbide

cerium oxide

Some of the uses of these grains and powders are as follows:

1. Abrasive grains are used in sandblasting, in grinding and polishing wheels, and in cut-off wheels.

2. Grains and powders are used in abrasive belts, sleeves, disks, rolls, bands, stones, sheets, and cones.

3. Grains and powders are used as abrasive media for barrel tumbling, polishing, buffing, lapping, honing, and vibration finishing.

Abrasive Grains

Aluminum Oxide

The abrasive grains most used in metal and plastic polishing are aluminum oxide, Turkish emery, silicon carbide, and diamond, both natural and synthetic. Some boron carbide is also used for polishing.

Aluminum oxide is the most widely used abrasive grain for metal and plastic polishing. The loose grain is used on set-up or polishing wheels and, in some instances, for lapping and barrel finishing. The grain is also used in bonded grinding wheels, abrasive belts, cut-off wheels, cartridge rolls, cones, sleeves, and in satin or greaseless compounds and barrel and vibratory polishing media.

Aluminum oxide grain as used today is manufactured from alumina, which comes from bauxite ore, and because bauxite comes from various parts of the world and contains different impurities, such as iron, the impurities in aluminum oxide grains also vary. This variety results in different moisture or water absorption ability, cutting quality, and color, which ranges from a light tan to a dark reddish brown; consequently, aluminum oxide grains will vary in price.

Aluminum oxide has a hardness of 9 on the Moh scale compared to a hardness of 10 for diamond. This hardness makes it an ideal abrasive for the polishing of ferrous and nonferrous metals. In use, aluminum oxide grain fractures with sharp edges, thereby continuing to cut well, making it economical to use. Aluminum oxide grains may be sized from as coarse as #8 to as fine as #400 and in some instances as fine as #1000, but grain sizes most commonly used range from a rough #36 grit to as fine as #240 or #320, which are used in the final polishing steps before a buffing operation.

Emery

There are two emeries available, both of which are natural. One emery is "American" and is mined in the United States; the other emery is "Turkish" and is mined in Turkey, with the ore being brought to the United States and ground here. Both emeries are black, with the American being somewhat dull in appearance, while the Turkish seems to be more "alive" or more light refractive. American emery is too soft to use for polishing, while Turkish emery, having a hardness approaching that of aluminum oxide, is more desirable.

Owing to chemical reasons, we do not use Turkish emery on non-ferrous metals or plastics: it seems to "drag" and overload with metal. However, Turkish emery is ideal for use on ferrous metals; and, while it is ground in grits from as coarse as #36 to as fine as #400, the largest usage and demand is for #180 and #220 grits as the final polishing steps prior to the buffing operation. While Turkish emery may and is sometimes used only for polishing steels, the demand has been so great and the supply has been so limited, that aluminum oxide is used. Turkish emery is favored because it fractures in such a manner that it does not score the metal as deeply as aluminum oxide and, consequently, it is easier to buff out the scratch lines.

Silicon Carbide

Silicon carbide is nearest to the diamond in hardness, having a hardness of 9; it is harder than aluminum oxide and is superior for the grinding and polishing of very hard metals. Silicon carbide is used in grits such as #60 and #80, and in the finer finishing grits for nonferrous metals. Silicon carbide is also used in abrasive belts and cut-off and grinding wheels and frequently with water or a light lubricant when used on plastics, ceramics, and glass. Because silicon carbide is more expensive than aluminum oxide, it is not extensively used on metals.

Diamond

Diamond, with a hardness of 10 on the Moh scale, is the hardest of all the abrasives. Flawed diamonds and chips, called "bort," are graded and sized and have many uses, in lapping, honing, grinding, and polishing. In addition, the flawed diamonds are used in cut-off wheels to cut very hard metals, such as tungsten carbide, and to cut concrete, stone, marble, glass, and ceramics.

Synthetic diamonds are being manufactured having a hardness almost the same as natural diamonds. These synthetic diamonds are used in much the same way as natural diamonds. In some instances, both the natural and synthetic diamonds are used in very fine meshes, with water or oil lubricants. Because diamond abrasives are expensive, methods have been developed for the reclamation of as much material as possible: there are companies specializing in such work.

Boron Carbide

Boron carbide is another very hard abrasive material that is not widely used, owing to its cost and to a lack of application knowledge.

Zirconium

Zirconium is often alloyed with other materials and used on belts for polishing. Owing to its cost and lack of application knowledge and experience, the acceptance of zirconium as a polishing abrasive has been slow.

If technical information regarding abrasives is desired, we refer you to *Abrasives* by L. Coes, Jr., Springer-Verlag, New York.

Abrasive Flours

In the manufacture of abrasives from raw ores to finished products, there are primary and secondary crushing or grinding operations followed by pulverizing. In the grinding operations the gangue or material judged to be worthless is disposed of by either a washing operation or by a stream of air blowing it away. The pulverizing operation or operations are done by machines in which the materials flow over screens or sieves that separate the ground material into the desired mesh sizes. The pulverizing operations generate dust that is collected by dust collectors, such dusts being called "dust collector fines" and being sold as such.

Alternatively, the pulverizing can be done by machines crushing the raw ore into powders that are screened or sieved into the various desired mesh sizes–these are referred to as "flours." Some of the "flours" used in the manufacture of the polishing/buffing/coloring compounds are aluminum oxide, various aluminas, clays, lime, silicas, and tripoli.

These abrasive flours are usually used in bar or liquid buffing compounds. Some of the flours may be used in tumbling, in vibratory equipment, and in some instances, in precision sandblasting of burrs or in dry precision polishing of very small and delicate parts such as parts in heart implants, jet engines, or electronic equipment.

The alumina flours in various mesh sizes are used in precision sandblasting and also in bar and liquid buffing compounds.

Aluminum oxide flours are used in certain sandblasting operations where the specifications may not be so rigid. While much of this flour is the waste material from the grinding of aluminum oxide grain, a portion is purposely ground and from which the iron and sulfur are removed, the main use being for the manufacture of buffing compounds for ferrous metals such as steel and stainless steel.

The iron oxide flours are used in rouge compounds for the buffing and coloring of metals and plastics. These oxides are in various soft to hard grades and very fine mesh sizes, and have a good grease absorbing quality; they make fine rouges. These rouges are used to provide scratch-free and highly lustrous finishes on gold, platinum, rhodium, and silver.

Tin oxide, cerium oxide, boron carbide, and diamond dust are used extensively for polishing rocks, marble, and exotic metals—as flour in conjunction with special oils. In addition, they may be used in sandblasting operations.

The tripolis and the silicas are used in buffing compounds, both liquid and bar, for use on nonferrous metals and plastics. These powders or flours are too soft for use in buffing ferrous metals but are ideal for the nonferrous metals and plastics: they do an excellent job without the deep scoring associated with the harder flours. In addition, these buffing compounds are usually cleaned easily and their cost per pound is lower.

Powdered or flour pumice is used with a brush or swab to clean bits of buffing compound, etc., that may have been trapped in cracks and crevices of parts. The largest user of this material is the silver plating shop or department.

We find diamond dust being used in conjunction with oil in the polishing of rocks by "rock hounds." Because the diamond is the hardest material known to man, it is a superb material for use in polishing the exotic or very hard metals and in polishing cut diamonds and other precious stones.

Abrasive Belts, Cartridge Rolls, Disks, and Sandpaper

Abrasive Belts

Abrasive belts are replacing polishing wheels, because belts are durable, fast cutting, very uniform on the surface or "face," and easily and quickly changed. They are available from the coarsest grits, #24, through as fine as #800 grit, in both a very flexible and a stiff cloth backing, in dry and waterproof material. The abrasive grains are of aluminum oxide, garnet, silicon carbide, and zirconium.

Belts with aluminum oxide grain are used on metals, both ferrous and nonferrous, and, in some instances, on some plastics; silicon carbide belts are used on both metals and plastics; and zirconium belts are primarily used on specialty metals. The abrasive belts may be purchased in widths as narrow as 1/4 in. or as wide as 110 in. and in lengths as long as 50 yards. It is suggested that you consult with your supplier as to the proper belt for your needs.

The equipment needed with abrasive belts depends on the production required, the type and shape of the work pieces, and the work to be done on the part. The auxiliary equipment includes a belt sander in an upright position or a horizontal position, possibly having a platen, an idler, and contact wheel; or automatic equipment may be called for.

Cartridge Rolls

The cartridge roll is made of an abrasive cloth rolled and glued simultaneously; the grit size is as desired. Cartridge rolls, 1/8 or 1/4 in. in arbor sizes, are straight or tapered, with a specified diameter, length, and arbor size, with or without a mandrel. Cartridge rolls are used in an electric drill to reach areas not accessible with polishing wheels or belts, i.e., corners, niches, or bowllike depressions, to remove light burrs and scratches.

Abrasive Disks

Abrasive disks are available in paper or cloth backing, in grits sizes from #24 to #600, in aluminum oxide or silicon carbide, in wet or dry materials, in diameters from 1/4 to 30 in., in a nonglue backing or a pressure-sensitive backing, in "J" light or "X" heavy material. We also have fiber disks—a heavy type of backing for rugged kinds of polishing on heavy flat or rounded metal surfaces.

Abrasive Wheels

For Grinding

Grinding wheels have abrasive grains from #24 to #1000 mesh, bonded together under high pressure for grinding applications ranging from use on a coarse casting to achieving a high-precision finish. They are also used to sharpen or hone to a very sharp edge.

For Polishing (Set-up Wheels)

A polishing wheel may be composed of either sections of sewed cotton, wool felt, or canvas buffs (sewing 1/8, 1/4, 3/8, and occasionally 1/2 in. apart).

Cotton wheels may be made of either full disk material or pieced cotton, with sections approximately 1/4 in. thick and with enough sections balanced and glued together to provide the thickness desired. These wheels are pressed in a wheel press while curing, usually over night; they are then checked for balance, faced, or have the face prepared for use.

Miscellaneous Abrasive Cloth Items

Flap wheels, which are a fairly recent innovation, are used in place of polishing wheels and where it is not possible to use abrasive belts.

Shop rolls are used by machine shops, mechanics, maintenance personnel; they are from 1 to 36 in. wide and 50 yards long, with grits from #24 through #800 mesh.

Contact Wheels

The contact wheel is the wheel on the polishing/buffing lathe over which an abrasive belt runs, from the contact wheel to and over an idlerpulley (Fig. 17). The contact wheel may also be a wheel that accepts an abrasive sleeve or cone. The contact on the wheel is the point where the part that is being ground or polished meets the abrasive belt, cone, or sleeve. The contact wheel may be composed of a number of buff sections stacked to make a face that is large enough for the width of the belt, cone, or sleeve being used. The contact wheel may also be a polishing wheel, a wool felt wheel, or a solid or serrated rubber-faced metal wheel. A contact wheel may vary in diameter from as small as 1 in. to as large as 30 in., and from a width as narrow as 1/4 in. to as wide as 130 in.

The contact wheel composed of the stacked buff sections is not as efficient as other wheels, but it offers the flexibility needed when working on intricately shaped parts. This type of wheel is inexpensive.

The wool felt wheel offers a more even or level face than the wheel composed of buff sections. This wheel is available in varying degrees of hardness, from "extra soft" to "rock hard," which also determines its flexibility. While the felt wheel is costly in terms of dollars, its uniformity of wear and shape retention can be economical in terms of long life. In addition, this wheel may be contoured or grooved as needed.

The rubber-faced metal contact wheel may be one with a "flat" face or one serrated or "grooved." This wheel is selected on the basis of diameter, face width, and hardness of the rubber: the serrated wheel being the most common. While the wheel is costly in terms of dollars, its great advantage is that under centrifugal force the serrations "hump" up, causing the belt to cut at the top of the "hump"—this speeds production and extends the useful life of the belt, which offsets the cost of the wheel.

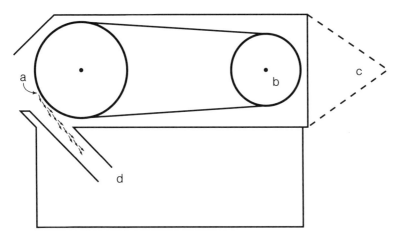

Figure 17. (a) Contact wheel working area; (b) idler pulley; (c) belt guard or hood, which has a hinged cover on the outside allowing for quick belt changes and is manufactured for left or right hand operation; and (d) exhaust outlet to system dust collector.

(The Chicago Rubber Co. through your supplier offers technical advice plus a choice of wheels to fit your need.)

The slotted or sectional contact wheel for the most part is used in small diameters and face widths from 1/2 to as wide as 10 in. with the diameters to 6 in. These wheels are available for use with abrasive sleeves or they are available in tapered or conical shapes for use with abrasive cones. The centrifugal force holds the bands or cones in place—changing is very quickly done. Such wheels are very useful for working on wood, plastic, and metal.

It is suggested that your supplier be contacted for his help in the selection of the best contact wheel for your need.

Probably the most commonly used contact wheel consists of a hub and replaceable tire, which is a durable serrated rubber surface; standard serrations may be ordered or the serration may be specified by the buyer. (Typical contact wheels are shown in Fig. 18.) With the removable hub, tires may vary in width, hardnesses, and design. Suggested sfm are 5000 to 7000. (Note. Excellent working diameters are 12 or 14 in. with a shaft speed of 1800 or 2000 rpm.) Some typical dimensions are shown in Table 1.

As seen in Fig. 18, the serrations on these small wheels are cut at an angle so that under the centrifugal force, the face of the wheel expands; this expansion holds the abrasive band or sleeve in place.

Many other kinds of serrations are available. (See Fig. 19.) The wheels

Table 1

Wheel Diameter	Band Length	Recommended Speed Range
3/4"	2 3/8	Approx. 10,000 to 12,000 rpm
1"	3 7/32	Approx. 10,000 to 12,000 rpm
1 1/2"	4 25/32	Approx. 6,000 to 10,000 rpm
2"	6 11/32	Approx. 4,500 to 7,500 rpm
2 1/2"	7 29/32	Approx. 4,500 to 7,000 rpm
3"	9 1/2	Approx. 3,600 to 7,000 rpm
3 1/2"	11 1/16	Approx. 3,600 to 7,000 rpm
4"	12 5/8	Approx. 3,600 to 6,000 rpm
5"	15 25/32	Approx. 3,000 to 5,000 rpm
6"	18 15/16	Approx. 3,000 to 4,000 rpm

Figure 18. Typical contact wheels.

as illustrated are manufactured by the Chicago Rubber Co., a leading manufacturer of serrated rubber contact wheels: check with your supplier or the manufacturer to be sure of selecting the proper wheel for your needs.

Selecting the correct rubber contact wheel can be as important as selecting the best abrasive belt. In general, hard rubber wheels will provide faster stock removal than softer wheels while leaving a coarser finish. Also, the centrifugal force from the speed of the wheel causes the surface of the wheel to harden: a medium hard, approximately 60 Shore wheel, is one that may generally be used for various grit size belts. A medium soft wheel, approximately 40 Shore, may be used on the softer metals, wood, and some plastics if used wet (with flowing water).

The following illustrations are of the different serrations available.

Figure 19. Typical contact wheel serrations: (a) cog; (b) x serration; (c) a specially molded serration available only in 14 in. diameter wheels of 2 or 3 in. widths.

Use of "Hot" or Animal Glue

The selection of the proper strength glue for use on set-up or polishing wheels is very important. A glue that is too weak will not hold up resulting in a short "life" for the set-up, and, conversely, a glue that is too strong results in a brittle set-up that quickly breaks off of the wheel.

The strength of glue is measured in grams—this strength varying from the low of 108 g to the high of 477 g: experience and usage over the years have shown that a strength of 251 g is recommended as having the best qualities for use on set-up or polishing wheels.

The following list is the recommended method of using animal or "hot" glue:

1. Put a measured quantity of water (cup, quart, etc.) into a glue pot. Then stir in a like amount of dry glue, stirring as you pour until the glue has absorbed all of the water, and continue stirring until the mixture becomes too thick to stir. (Caution: NEVER pour the water into the glue: the glue absorbs the water too rapidly to become completely wet.)

2. Allow the mixture to soak for a minimum of 1 hr; for the best results, wait longer or overnight. Next, heat the mixture in a thermostatically controlled electric glue pot: the glue will melt in 20–30 min. (Note: If the glue is to be heated over a gas burner, use a double boiler and heat over the lowest flame possible   glue heated over 150°F will burn, losing its strength in proportion to each degree over 150°F.)

3. Using a round glue brush, free of any grease or oil, apply a thin coat of glue to the wheel. Allow the glue to become slightly tacky or sticky to the touch, then apply a second thin coat. When the second coat becomes slightly tacky, and with a rod or piece of pipe through the wheel arbor, roll the wheel in the particular abrasive grain to be used, lightly pounding or bouncing the wheel in the grain as you roll

it; this ensures that the wheel surface absorbs the grain. (Note: If various mesh size grains are to be set-up, start with the finest grain to be used. When finished, be sure to wash all of the grain out of the brush, washing or rinsing until the water runs clear. This preserves the brush.)

4. Allow the wheel to cure overnight in a reasonably moisture-free atmosphere. If an oven is used, do not heat higher than 110°F and have a tray of water in the oven; this prevents the wheel surface from becoming too brittle. Or, as some do, use the hull of an old electric refrigerator with a 40 watt bulb burning; this bulb provides sufficient heat to cure the wheel. Also, have a small tray of water in the bottom.

5. Before placing the set-up wheel on the shaft of the lathe, crack the surface of the wheel with a piece of metal; then place the wheel on the shaft, and, using a piece of broken grinding wheel or a piece of metal, apply it lightly to the face of the wheel, to knock off any high spots. This will true the face. The wheel is now ready for use.

Polishing Wheel Cement

Polishing wheel cement—sometimes referred to as "cold glue"—is a mixture of water, sodium silicate, and various chemicals, which make a binding material for use on set-up or polishing wheels. This material has a high melting point after curing and strong adhesive qualities; thus it is much stronger and more heat resistant than animal glue and an ideal material as a binder of abrasive grain to the polishing wheel.

The following list explains how to use polishing wheel cement:

1. Be sure that the wheel is free of any grease or oil. Use a round glue brush also free of any grease or oil; apply a thin coat of cement to the wheel and allow the cement to become "tacky" (approximately 5 min.), then apply a second thin coat of cement. Allow this second coat to become "tacky" or "sticky" to the touch. (Note: Sometimes it may be best to apply only one coat, since in the two coat set-up, should the first coat become too dry or set before applying the second coat, the "set-up" will not adhere to the wheel.)

2. Once the wheel is ready, insert a rod or piece or pipe into the arbor of the wheel, then roll the wheel in the grain you plan to use, lightly pounding or bouncing the wheel as you roll it—this action more firmly embeds the grain into the wheel surface.

3. Following the "set-up," the wheel may be allowed to "cure" by setting over night, or the wheel may be cured in an oven or heated room. Do not heat above 110°F. Also have a pan of water sitting in the room or oven—this provides that necessary bit of humidity which keeps the wheel "set-up" from becoming too brittle. (A helpful suggestion: An old refrigerator casing or hull with a 40 watt light bulb and a small pan of water in it make an ideal curing arrangement for a small lot of wheels.)

4. If the plan is to set-up a number of wheels in various grit sizes of grain, set-up the wheels of the finest mesh size, and follow each additional grain size in succession; doing this avoids the need to clean the glue brush after each set-up. This procedure also prevents the mixing of a coarser grain with the finer grain. (This assumes that only the one glue brush is being used; using the one glue brush also avoids the potential mix ups that can occur where there are many brushes.)

5. Since the cement is water soluble, be sure to rinse out the brush thoroughly in running water—continue the rinsing until the water runs clear. If this is not done, you may end up with a brush "set-up" so solid that it is unusable.

6. After the wheel has cured, take a piece of pipe and crack the surface of the set-up (the face of the wheel). Following this, mount the wheel on the shaft of the lathe, take a piece of pipe, or a piece of a broken grinding wheel, or a brick, and, as the wheel spins on the shaft, lightly apply it across the face; this removes any high spots and trues the surface—you thus have an even wheel-face for polishing.

7. Once the set-up or wheel face has worn to the point where it does not cut—requiring the wheel to be set-up again—clean the wheel of dirt and grease by knocking off any chunks of the old set-up with a piece of old grinding wheel, or a wheel rake, following with a piece of pumice stone. (This procedure lightly shreds the wheel-face and provides a surface for the reapplication of cement.)

Greaseless Compounds

Greaseless compounds are mixtures of glue, abrasive grains (aluminum oxide), and certain chemicals that both dry and harden. These compounds are available in #80 to #400 grit.

Greaseless compounds may be used on almost any kind of a buffing wheel—loose or sewed cotton buff, string wheel, or felt wheel. These compounds may be used for light deburring to obtain a "satin" or grained finish, for the removal of light scratches or imperfections in metal, and on nonferrous and ferrous metals.

To set-up or prepare a "satin" or "Lea" wheel (the compound is packaged in either a paper, plastic, or metal tube):

1. The wheel should be turning at approximately 5000 sfm.

2. Peel the tube back approximately 2 in. Apply or hold the compound against the wheel and turn off the power. The glue will cause the wheel to grab the compound, so have the wheel running, apply compound, turn power off—do this until a "head" or layer of compound has built up. Note: If your power to the wheel is not too strong, you may find that you are able to build up the "head" without having to alternate the stopping and starting.

3. As this compound air dries, you may leave the wheel running for a few minutes, and then remove it and set up additional wheels. It is very important to allow the wheel to "cure" or dry for at least 1 hr before using, and for better results and longer life of the "head," allow it to cure over night. Also, by preparing a number of wheels at the same time, production may continue with the only interruption being to change wheels.

4. Mark the wheel with an arrow to indicate the direction it turns as you use it; if the wheel is used in the wrong direction, the "head" will last only a few minutes.

5. Whether you are deburring, polishing, high lighting, or "satin" finishing, oscillate the part from side to side—this not only causes the wheel to wear evenly but avoids lap marks, skips, and allows for an even pattern to the finish.

A recent development is the "Scotch-brite" wheel, sheet, or disk as an alternate to the "satin" wheel. These wheels do an excellent job when used correctly. However, they are expensive, not only in price but in the life of the wheel, since they do not last long.

See Chapter 12 for another alternative to the greaseless compound: "Kool-Kut."

Kool-Kut

Kool-Kut, manufactured by Siefen Compounds, Inc., Wyandotte, MI, is a liquid cement (a cold glue) containing aluminum oxide grain; it is available in grits from #60 mesh to #500 mesh. When cured, this material has a melting point of 1200°F. Since most greaseless compounds are composed of animal glue and aluminum oxide grains and have a melting point of less than 300°F, the high melting point of the Kool-Kut saves both time and money when compared with other greaseless compounds.

Owing to its high melting point, wheels set-up with Kool-Kut have many uses: as a polishing wheel, for deburring or removal of flash from castings, to provide a "satin" or grained finish, to remove deep scratches or imperfections in metals, to "highlight" oxidized finishes. The following list explains how to best use Kool-Kut.

1. Use a 100% cotton wheel free of oil and grease. Do not use synthetic material—the Kool-Kut does not adhere well to it.

2. Since the grain in Kool-Kut will settle, be sure that the grain is thoroughly stirred into suspension. (A paint-shaker is superb for this use.)

3. Use a clean brush—preferably a round glue brush—to apply a coating of Kool-Kut to the face of the wheel. Caution: Do not try to apply more than the one coating.

4. Allow the wheel to cure a minimum of 1 hr preferably under a 60 watt light bulb. If possible, it is best to cure overnight. (A 50 or 60 watt bulb in the shell of an old refrigerator is excellent for this purpose; have a pan of water in it to avoid a too brittle cure.)

5. When cured, mount wheel on the machine shaft; then, use a piece of metal, a broken piece of a grinding wheel, or piece of brick—almost anything—take whatever you have and lightly pass it across the face of the wheel to knock off any high spots, thus trueing the face.

6. Carefully use the fingers to break the surface of the Kool-Kut—the finer the breaks, the sharper and more uniform the finish.

7. To be most effective and to provide a better finish, oscillate the wheel from side to side in operation. Do not use an up and down motion, since this motion distorts the finish.

8. If the run is to be a continuous production, it is suggested that a number of wheels be prepared. To avoid having a large number of brushes or cleaning the brush after each grit size is set-up, start with the finish grit and then set-up each coarser grain. Mark the grit size on each wheel to avoid any confusion.

9. As previously noted, wheels may be cooled or cured in various ways—air cured, oven cured, etc.

10. Important: After the wheels have been set-up, wash the brush or brushes in running water, until the water runs clear. If you do not, you will have a cement hard and unusable brush.

11. Very important: After using the Kool-Kut, tightly replace the lid on the container. And, if it is not used daily, turn the container upside down weekly, since the grain will settle. This turning will make the Kool-Kut easier to stir and mix the next time it is used.

12. Additional information: Water is the solvent for this material. When adding water, add only a minute amount, since the Kool-Kut thins very quickly. While we advised against applying more than one coating, with care and due attention, additional coats may be applied by allowing each coat to become sticky before applying another coat. If a coat is allowed to get too dry for another coating, the added coatings will only fly off.

Buffing Compounds

Description

In metal and plastic finishing, your most expensive item is time. Time as applied to the number of parts or pieces per hour, the amount of compound used per hour, the wear and tear on the buffing/polishing wheel per hour. The major determinant of these costs is the capability or quality of the bar or liquid compounds and buffs being used. Prices are or should be secondary!

Since the buffing compound business is highly competitive, the manufacturer may use low-cost filler materials in both bar and liquid compounds to both increase weight and lower the overall cost per pound. By using a given amount of compound and checking the number of parts finished with that compound, the time required to buff those parts, a true cost may be calculated. Since these compounds are formulated with various greases and binders, the cost of cleaning parts also becomes important.

The types of compounds used are tripolis, steel and stainless steel compounds, plastic compounds, rouges, plus many specialty compounds; these various compounds may be purchased in either bar or liquid form.

Types and Uses of Buffing Compounds

Tripoli Compounds

Tripoli compound in bar or liquid form is a reddish brown or light tan material. There are many variations in the quality of the abrasiveness of the tripoli compounds, varying from a "coloring" to a "fast cut" tripoli and from a "dry" compound to a very "wet" one.

Tripoli is used on nonferrous metals and some plastics. The choice of the tripoli to use may be based on the type of metal and the condition of the part: whether it has a rough or a smooth surface, it is a casting or sheet metal, etc. Also, important considerations include the finish needed or desired, and, whether the part is to be coated with a clear material, painted, or electroplated.

A bar of tripoli compound 10 × 2 × 2 in. in size will weigh on the average approximately 2 1/4 to 2 1/2 lb., while a gallon of liquid tripoli will average approximately 8 lb.

It is suggested that samples be tested before purchasing any quantity.

Steel Compounds

Bar or liquid steel compounds vary from an "almost" black or dark gray to a light gray in color, the variations in color being due to the composition. These compounds are used on iron, various grades of steel, stainless steels, and certain aluminum or brass castings. As in the other compounds, the quality of the abrasiveness varies according to its use on the metal being buffed—varying from the sharpest or most aggressive to the mildest or "coloring," compound.

It is important to note that sisal buffs should always be used when buffing ferrous metals; about the only occasion for the use of cloth buffs would be as a means of improving the fineness of and increasing the brightness of the finish.

The bar steel compounds of approximately 10 × 2 × 2 in. will weigh approximately 3 1/2 lb, a gallon of liquid steel compound will weigh approximately 13 lb. These compounds are a mixture of various oils and greases plus an abrasive "flour" of aluminum oxides, and, possibly, a silica as a low-cost filler.

Steel compounds will vary in "cut" depending on the kind of sisal buff being used and the amount of grease in the compound, whether it is a liquid or bar compound. It is suggested that samples of both the buffs and compounds be tested for the best results. Your supplier can assist you.

Stainless steels vary from the impossible to buff to the easily buffed. We find that stainless steel "likes" a "dry" or "medium dry" compound on either an untreated full sisal or laminated sisal buff. And, as a suggestion, you may find that the brilliance of the finish can be increased by the use of that same compound on a cotton buff—loose or 3/8 sewed (speed about 6000 rpm).

White Diamond

White diamond, usually a light beige colored material, is a compound formulated in bar and liquid form of special oils and greases and silicas as the abrasive. It is greatly prized by the jewelry manufacturing industry for use on precious metals. White diamond compound is used primarily on nonferrous metals, particularly pewter and die-cast metal, to give a bright clean finish. While the compound is made in various grades or qualities, its most popular form is a fine mesh silica and very special greases in bar form.

Because the quality of white diamond may vary greatly from manufacturer to manufacturer, we suggest that very careful testing be done before any quantity purchase.

Chrome Rouge

Chrome rouges vary greatly in quality and use. These compounds may be in bar or liquid form and have a fine sharp cut varying to a very mild cut. These rouges are composed of various mild oils and greases with an alumina as the abrasive and are manufactured in several grades for different uses.

The chrome rouges may be used on ferrous and nonferrous metals, and also on many plastics; therefore, it is best to test for the rouge best suited to your needs.

Jeweler's or Red Rouge

Jeweler's or red rouge is manufactured with the finest oils and greases and has as the abrasive very fine iron oxides; it is available in either bar or liquid form. This rouge is superb for bringing out the deep richness and beauty of the color in the nonferrous metals, particularly brass, gold, platinum, and silver. These rouges usually present no problem in cleaning. We do suggest, however, that testing be done before any quantity purchase.

Emery Cake

Emery cake is a compound composed of silica, an abrasive grain (either aluminum oxide or Turkish emery) and the necessary greases and oils; it is colored with lampblack. Many years ago emery cake was used with Tampico wheels or brushes for buffing cast iron and steel. Today, emery cake is usually used as a lubricant on set-up or polishing wheels. (It is still used with Tampico wheels in some special situations.) Emery cake is manufactured mostly in #180 or #220 grit, in bar form or usually poured into tubes. Using emery cake on polishing wheels helps to improve the cut and, at the same time, prolongs the life of the set-up. While this is a dirty compound in use, it is very easily cleaned.

Crocus

Crocus is probably the least known of buffing compounds, yet it is available in both bar and liquid forms. This compound is made of special oils and greases with a low cost iron oxide as the abrasive. Usually the iron oxide used has a fairly good "cut": there are a few manufacturers using crocus to buff mild or "soft" brass (sheet and castings) pewter and diecast.

Fiberglass Compound

Fiberglass compound is light beige in color, being formulated of special oils and having a silica as the abrasive. Because fiberglass is easily "burned," the compound used must be oily or greasy enough to avoid this. Since fiberglass is laminated in layers to form molds, skis, boats, and many other items, the compound should also be fine enough to provide the finish desired and, at the same time, easily cleaned. While it may seem easy to get a good compound, it is best to do extensive testing of the compounds from various manufacturers, since using a poor compound can be very expensive.

Synthetic Marble Compound

"Cultured" or synthetic marble compound is a mixture of special resins and solids. Because synthetic marble is a somewhat tricky material to buff in that as it cures, it hardens. This curing is subject to the manufacturing methods used, the mixture of the various ingredients, the climate at the time of manufacture, and so on. While a bar or liquid compound is available from the supplier, it is suggested that a program of extensive testing of compounds be established; we have found that so doing is profitable.

Some manufacturers have formulated compounds specifically for use on cultured marble. While such a compound in liquid form is available, a bar compound will be found to be much cleaner and easier to use because of the horizontal nature of its application; it also provides a much more lustrous and scratch-free finish.

Satin Finish Compound

The "satin finish" or greaseless compound is a mixture of an abrasive (aluminum oxide) and glue containing hardening and drying chemicals. Since this compound hardens when exposed to air, it is packaged in either aluminum, plastic, or airtight paper tubes. Since these compounds do vary in quality, it is suggested that testing be done before quantity buying; the variance is usually in the ease of applying it to the wheel and how well the compound holds up. This compound is available in grit sizes running from as coarse as #60 grit to as fine as #320 grit—it is used for light grinding, for deburring, for the removal of slight metal imperfections, for highlighting of an oxidized finish, and for providing a "satin" or grained finish on metals. "Satin" compounds may be used on both ferrous and nonferrous metals, also on some plastics. This compound also leaves a bright, dry, and clean finish.

Liquid Buffing Compounds

Liquid buffing compounds are primarily used with automatic machines for volume production; however, they are used also with semiautomatic machines as well as with some hand buffing operations. With proper and efficient equipment set-up, liquid compounds can be a low-cost and productive way to maximize production of ferrous and nonferrous metals, and, additionally, some plastics. Since these compounds are also composed of water-based emulsions, they present no problems in the cleaning of parts that have been buffed.

The quality of liquid compounds may vary from manufacturer to manufacturer; this quality depends on the percentage of liquid (water and oils or greases) used and the quality or type of abrasives. It is suggested that comparative tests be made to determine the most economical compound to use.

Liquid buffing compounds are applied by spray guns operated by timers that control the air to the gun. The compound is pumped from a drum. Figure 20 is a schematic showing a suggested layout for either automatic or hand operation. The following sections on the proper use of liquid compounds is courtesy of Siefen Compounds, Inc.

Modern Method of Spraying Liquid Buffing and Polishing Compounds

The spraying of liquid buffing and polishing compounds is relatively simple. The necessary equipment consists of (1) a spray gun, (2) a means of controlling the air to the gun, and (3) a source of compound under controlled pressure at the gun.

There are many spray guns of various manufacturers used for spraying the buffing compound. Many of these are paint spray guns or adaptations of them. All work satisfactorily with the proper air control; their oper-

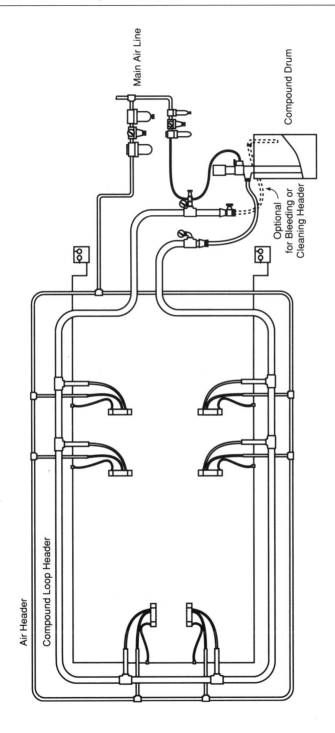

Figure 20. Schematic layout of typical installation to utilize liquid buffing compounds.

ation varies with different manufacturers. The guns fall into two broad classifications: the one air-hose gun and the two air-hose gun.

The one air-hose gun consists of fewer parts and is therefore simpler to operate—it functions by controlling the air at or to the gun. Part of the air that passes through this one air hose to the gun is used to lift the needle by pressure against a piston or diaphragm. The remaining air passes on through to the front of the gun and is used for atomizing and shaping the spread of the compound. When the air is shut off, the needle is seated by spraying pressure and thus stops the flow of the compound.

The two air-hose gun is controlled by various methods. Generally one air hose supplies air at high pressure to move a piston that lifts a needle. The second air hose furnishes air at a lower controlled pressure for atomizing the compound. This type of gun may require a complicated valve for controlling both air lines, or it may have built-in features that require only the control of the high pressure line. When air at the proper pressure is supplied to a properly assembled spray gun, the needle will lift, and, if compound under pressure is present at the fluid tip, the compound will be sprayed.

Placement of Guns

The placement of the spray gun is extremely important. Many people disregard the waste that occurs when the gun is placed too far away from the buff or the work. The ideal setup is to have the gun spray the work. Generally, this procedure will not require any adjustment, as happens frequently when the buff is sprayed instead. The next best location is between the buff and the work. The least desirable procedure is to spray the compound onto the buff, unless the gun can be placed very close and kept there at all times in order to avoid blowing away the compound.

In most instances, except for indexing-type machines where the work surface is large, careful study is required in order to find the most efficient way to spray the work. If the work cannot be sprayed, try to spray as close to the intersection of the work and buff, on the side where the buff goes into or hits the work. Compound will be saved and you will be assured of clean work.

If the buff only must be sprayed, keep the gun as close to the buff as possible. Some guns are made with an adjustable spray width. Other guns have set air caps for a set width, which generally will spray a width equal to the distance away from the buff. For instance, if the gun is 3 in. away from the wheel, the spray will be 3 in. wide. This width varies with different manufacturers. In this event, some means must be provided to adjust the gun to compensate for wheel wear, otherwise compound will be wasted and the quality of the work will decrease since there will not be sufficient compound applied to the buff to give maximum cut. The farther the gun is away from the wheel, the more compound will be blown away by the convective air produced by the high speed of the buff.

For the same reason, it is not advisable to use air pressure that is too high at the gun. If low pressure is used, the compound will be sprayed in droplets rather than misted, thus ensuring that more compound will be placed on the buff.

Be sure that the entire working area of the buff is sprayed with the proper amount of compound, because it will not spread from one part of the buff to another. However, if excessive amounts of compound are applied, some will be carried to adjacent parts of the buff but only in smaller amounts than is needed or necessary for good buffing.

When wide buffing wheels are used, the gun should be moved back and forth across the face of the buff or work. This saves the compound and buffs and produces more uniform quality of work as the coloring or cutting head on the buff is maintained correctly. We have seen many installations where a gang of three or more guns is used to spray a wide wheel. The waste is appalling and the machines are very dirty. This situation can be avoided. Proper placement of the spray guns and proper adjustment for wheel wear, when necessary, will pay for itself.

Wearing of Gun Parts

Many attempts are being made to reduce the wear on needles, fluid tips, and packings. Carboloy or rubber inserts are the favorite methods. The best way to keep wear to a minimum is to reduce compound pressure.

In a balanced system, where air and compound control are proper, the wear on parts is negligible. The use of a nonadjustable spray gun forces proper air and compound control, and we cannot recommend too strongly that this type of gun be used. This removes the possibility of operators adjusting the gun to suit themselves. The nonadjustable gun permits the needle to rise completely from the seat of the fluid tip and creates a large passage for the compound. Many months of wear from each needle and fluid tip is ensured.

Air Control

Foot valves, cam valves, and mechanical time valves were used in the early days of spraying to control the air that operated the spray gun. Electric timer valves, which operate solenoid valves, are now being used successfully for this purpose. Control is at its best because slight variations may be made instantly, which results in compound and buff savings and guarantees more uniform quality of work.

One example of a timer for most installations is the Siefen Compounds, Inc., Electronic Timer. This timer can be adjusted to allow a variable spray and a nonvariable spray time in intervals as short as 1/100 sec. In most instances, the spraytime is set for 1/10–1 sec. and kept at this setting. If less compound is needed, nonspray time can be increased.

Experience has shown that for average work on a 4-in.-wide wheel, the timer can be set for 1/10 sec "on" and 2 sec "off." Note: the shorter the on-spray time, the more economical the operation.

The "hills and valleys" of compound application are thus leveled off when timers are used. Electric timers may be mounted at any convenient location. The timer and the solenoid valve, which is part of the spray gun, are connected by a 115 V line. For better control, keep the number of solenoid valves operated by any one timer to a minimum. Over time, it is more economical to have a one on one approach for each wheel. This setup will give exceptional control and save compound and buffs to such an extent that the equipment will soon pay for itself.

All air used for spraying must be clean and oiled. Use a filter if you are not sure of clean air. A small amount of water and oil in the lines is not harmful to the compound. Install an oiler and be sure it is operating correctly. All spray guns, valves, and pumps require oil in small quantities to function properly and to keep wear to a minimum.

Air pressure regulators should be large enough so as to not bottleneck the air to and from them. Do not use a regulator smaller than the piping on either side. The volume of air requirements will vary with the size of the holes in the air cap and the interval of time the gun is spraying. A gun with no holes in the air cap other than the central atomizing hole will use approximately 2 ft^3/min at 40 psi. Guns with several large holes, used for spraying compound, will use as much as 15 ft^3/min. Since the spray gun is not operating at all times, be sure to find out the spray interval before calculating the consumption of air per minute.

Compound Supply

Buffing compound is supplied to the spray gun under pressure. A pump of the reciprocating type made for pumping grease is used in most plants. These pumps are subject to wear and will give satisfactory service only if operated slowly.

Most pumps are capable of pumping many times more compound than required. Since each gun requires only a small amount of compound, the pump should pump slowly and then only part of the time. Pumps will only pump when compound pressure drops in the compound line.

The best way to control the pressure of the pump is with a pump control valve, which is a diaphragm-operated valve that controls air to the pump through pressure in the compound line. When the pressure drops, air is fed to the pump and the compound pressure is returned to normal. Other means have been used, but they are not satisfactory.

Keep compound pressure as low as possible to prevent waste. Generally 2–10 psi *at the gun* is sufficient. Use as large a compound line as is possible. If in doubt about the size of the pipe, use a larger size rather than a smaller one. Use large drop lines, generally 3/4 in., to within easy reaching distance of the operator, and a fluid cock for shut-off where the hose is attached.

Use as short a hose as possible and do not coil it—always cut it to a suitable length.

Do not bottleneck the compound! Keep the compound lines the *same size*. Use a short hose from the pump to the line. Install a valve at the end of the compound line, which is shut off all the time. Provide a hose from this valve so that the line may be drained when necessary or when cleaned.

In order to tell if the compound is at the proper pressure, install a compound pressure gauge at each end of the line.

Do not contaminate the compound with lint, floor compound, etc. Provide a suitable cover for the top of the compound drum in order to keep it clean. Foreign matter, if small, will end up at spray gun and stop the flow of compound. This leads to fire!

A screen similar to a steam strainer should be placed at the beginning of the compound line and cleaned at regular intervals. Finally, economy can be obtained by using properly controlled oiled air, proper compound pressure, and accurate timing of the spray gun, which in turn will lead to more uniform, high quality, clean work and clean machines.

Liquid Compounds Daily Troubleshooting Guide

1. Remove all buffing dirt and foreign matter from drum covers and areas where it could contaminate the pressure centers.

2. Check the mesh strainer and compound line at the strainer to ensure compound passage.

3. Check the spray pattern and set it properly for each gun. Take a piece of cardboard and cover the area where the spray strikes the wheel. Spray once and check the pattern on the cardboard!

4. If a gun fails to spray, take the following steps:
   a. Make sure the gun is getting air, checking for breaks in the air hose, for a stuck solenoid, and so on.

   b. If there is air, remove the compound hoses at the gun. If compound is not coming through, shut off the pet cock at the main line, remove the hose, and reverse blow it with compressed air to remove all compound.

   c. If there is still no compound, open the pet cock on the main line. Make sure an adequate supply of compound is available.

   d. If there is no compound at the pet cock, the main line must be plugged. Use hot (160°F) soapy water to flush it. This should dislodge any object in the main line.

5. Now if there is compound and air to the gun, and no spray, inspect the gun as follows:

   a. Remove the aircap and nozzle.

   b. Open the pet cock to check the flow into the gun.

   c. Insert the end of a paper clip into the orifice outlet to push out all grains or others matter.

   d. Reverse blow the gun tip with compressed air.

   e. Replace all gun parts and turn on the gun to check for spray. If the gun sprays, you are back in business. If not, rebuild or replace the gun.

Liquid Buffing Compounds
Troubleshooting Guide
for All Automatic and Semiautomatic Spray Guns*

Problem	Solutions
1. No cut	a. Check compound application. b. Are guns clear and spraying? c. Increase compound pressure. d. Increase spray time. e. Increase dwell time.
2. Compound left on part	a. Check compound application—is there too much? b. Reduce compound pressure. c. Reduce wheel pressure. d. Decrease onspray time.
3. Light buff? Poor quality?	a. Check temperature of buff—is it too hot? b. Check wheel settings. c. Are guns spraying? d. Are wheels worn out?

* Courtesy of Siefen Compounds Inc., Wyandotte, MI.

e. Slow machine as a last resort.

4. Buffs slipping

a. Do flanges have keys in them?
b. Spacer and arbor holes correct?
c. Has nut bottomed out on thread?
d. Reduce buff pressure.

5. Inconsistent finish

a. Are parts slipping on the fixture?
b. Are fixture heights consistent?
c. Check for air in compound line.
d. Check for contaminants in com- pound line.
e. Is the on and off time on the compound timer set incorrectly?
f. Is pressure on the compound line high enough?
g. Is pump operating properly?

6. Poor buff life?

a. Reduce buff pressure to "QUAL- ITY BREAK POINT."
b. Are buffs getting too hot?
c. Check desired amperage draw "PREDETERMINED."
d. Is compound application prop- erly set: timers, pressures, clean filters, pump running?

7. Buffs overloading with compound

a. Reduce the "on" time at the timer.
b. Set buff pressure higher (more amps).
c. Make sure buff is not slipping on the arbor (shaft).

NOTE: THE BUILD UP OF THE BUFFING "HEAD" OF COMPOUND ON THE WHEEL SHOULD BE DAMP AND SOGGY.

8. Wheel ragging or tearing

a. Too much wheel pressure.
b. Wheel running in the wrong direction.
c. Wrong buff for the part.
d. Not spaced properly.
e. Buff out of balance.
f. Not broken in properly, by in- creasing timers and pressures on the compound.

NOTE: WHENEVER POSSIBLE, DO NOT OVER-PRESSURE THE BUFF BEFORE A GOOD AMOUNT OF COMPOUND HAS BEEN APPLIED TO THE FACE OF THE BUFF AND GRADUALLY LOWERED TO THE PART, WHICH WILL HELP LAY ALL OF THE THREADS DOWN ON THE BUFF FACE. THIS WILL BEGIN TO BUILD UP THE DESIRED CUTTING FACE.

Cleaning

Cleaning Polishing and Buffing Wheels

Polishing and buffing wheels become very dirty from the accumulation of removed metal, grease, and buffing compounds. Because the nonferrous metals are softer than ferrous metals, wheels will become loaded with debris more quickly, and, if not cleaned occasionally, will become so dirty as to be practically useless: this is true of both polishing and buffing wheels.

The surface or set-up wheel may be reasonably cleaned by the passage of a piece of pumice stone lightly across the face of the wheel. Or if the set-up has worn down to the extent that it needs to be set-up again, the surface of the wheel with its bits of grain and accumulated dirt and grease may be cleaned by the use of a carborundum stone first, then followed by a judicious use of a piece of pumice stone: this will remove any lingering debris and, at the same time, "true" the face of the wheel.

When cleaning buffing wheels, if they are heavily loaded with dirt, grease, and buffing compound, first lightly use a buff rake to knock out any lumps of accumulated materials, followed by the use of a piece of pumice stone.

Caution: The teeth of a buff rake tear the cloth face of the wheel, while the pumice stone frays. Do not use the buff rake too freely.

Note: An old red brick makes an excellent item to knock off the abrasive from a polishing wheel. It is also good for "truing" the face of the newly set-up wheel prior to polishing. Sometimes the buffing wheel will fray or throw outward some threads; carefully use the cutting edge of a knife to cut these off while the buff is running.

Cleaning of Bar Buffing Compounds

Bar buffing compounds are made of a combination of selected greases and oils, and combined with various dry materials to provide specific finishes on metal or plastic. These greases and oils may be animal or vegetable or a mixture of animal, mineral, and vegetable compounds. These compounds, when used for buffing, will leave a deposit. These compounds may be highly saponifiable and thus easily cleaned. (The waste from the buffing is composed of dirt, grease and abrasive material and metal removed by the buffing action.)

Many methods of cleaning are used; the most common cleaning method is vapor degreasing, using trichlorethylene or perchlorethylene or Freon. This method is very effective and not too costly. The trichlor will remove animal and vegetable oils, perchlor removes mineral and animal oils, and Freon either. Other methods of cleaning are hot soak cleaner, electrolytic cleaner, alkaline cleaner, Stoddard solvent, evaporative cleaners, and wiping with whiting, hot water, and a soap.

The buffing compounds used should be chosen with due regard to the method of cleaning available. If parts are to be electroplated, lacquered, coated or painted, it becomes very important that the cleaning is complete—proper cleaning minimizes rejects. Finally, the selection of a method of cleaning buffed parts is subject to various governmental regulations—be sure to check these.

Cleaning of Metal Parts

Parts or pieces that have been ground, machined, polished, or buffed must be cleaned prior to being coated or electroplated, or they must just be cleaned. These parts have oil, grease, dirt, bits of metal, abrasive grain, or buffing compound on them.

Cleaning may be done in one of many ways:

1. Vapor degreasing by
   a. Freon
   b. Trichlorethylene
   c. Perchlorethylene

2. Electrolytical cleaning

3. Hot caustic solution

4. Hot soak (soap) cleaner

5. Washing in kerosene or Stoddard solvent (the best results are obtained if the cleaning is done while the parts are still warm from buffing)

6. Lacquer and paint thinner are NOT recommended because they are VERY DANGEROUS to use

7. An alcohol

8. Wiping with whiting

The method of cleaning used may be determined by the kind of metal to be cleaned, the quantity of parts, the production required, and the facilities or equipment available. There also are government restrictions and conditions that must be adhered to. A brief description of the various methods follows:

1. Freon cleaning is a relative new usage of this material. While it requires special equipment, it is very efficient.

2. Trichlorethylene is used as a vapor cleaner of oils, greases, and buffing compounds. It is VERY TOXIC, and prohibited in many localities. It is an effective cleaner of animal and vegetable fats.

3. Perchlorethylene is also used as a vapor cleaner of oils, greases, and buffing compounds. It is also toxic and prohibited in some areas. It is a very effective cleaner of mineral oils and animal fats.

4. Electrolytical cleaning uses a solution in which the part to be cleaned is immersed and becomes the anode, while the solution is the cathode: it is an efficient and low-cost method. See your supplier for details.

5. Hot caustic solution cleaning is limited in its application. See your supplier for the necessary information about both materials and necessary equipment.

6. Hot soak or hot soap solution—again, we suggest contacting your supplier.

7. While kerosene or Stoddard solvent is used for cleaning buffed parts, neither material is particularly recommended. The use of these materials requires rubber gloves and mask, probably a brush, and then the results are often not too good. There is also the problem of the disposal of the dirty residue. We suggest that this may be a temporary solution until a better method is instituted.

8. Using lacquer or paint thinner is very much a fire and health hazard. While these materials are used, they are not recommended—their use is usually limited to a small number of parts. Their use requires masks and plenty of ventilation.

9. Alcohol, particularly isopropol, while costly, is an excellent cleaner for very small parts. Frequently, it is used in conjunction with ultrasound methods. We suggest contacting your supplier for details.

10. Whiting—a nonabrasive and extremely soft material—is excellent with which to wipe parts, since it absorbs grease and dirt well and offers a method of cleaning a limited number of parts when other means are not readily available. Note—a word of caution—do NOT use shop towels or any synthetic materials to wipe parts, they will scratch the buffed parts: Use cotton cloth that has been laundered many times, very soft cotton cheesecloth, or light double nap cotton flannel cloth.

When selecting buffing compounds, avoid those compounds containing paraffin; also be sure the compounds are highly saponifiable.

Pumice

Pumice is a porous and fairly soft volcanic material that is ground into various fine mesh sizes. It is used as an abrasive, primarily as a cleaning and mild polishing material.

Very finely ground pumice is used with water and a long-handled plater's brush to clean stubborn spots of grease or buffing compound off parts. It may also be used with a small flow of water onto a buffing wheel to provide a matte finish on a nickel- or silver-plated metal part. Such usage may be by the jewelry or electroplating industry.

Rock or lump pumice is used in the metal and plastic polishing and buffing industry to clean polishing and grinding wheels of accumulated grease, dirt, and buffing compound.

The metal removed when buffing nonferrous metals combines with the buffing compound to load the wheel, and if the wheel is not occasionally cleaned with a piece of pumice, the wheel becomes very ineffective and nonproductive.

While there are many sources for pumice stone, there are two main sources: Mount Vesuvius in Italy, and, in the United States, Levining, California. The pumice from Italy is the best quality, while the Californian pumice is a cheaper source.

Whiting (Calcium Carbonate)

Whiting is an inexpensive, very soft, and nonabrasive material derived from limestone. This material, owing to this softness, has many uses:

1. When used on the hands, whiting absorbs perspiration and prevents fingerprints on the workpiece.

2. Use frequently washed cotton rags, old sheeting, cheesecloth, or double nap flannel that has been dipped into the whiting to wipe off grease and dirt from freshly buffed parts.

3. Mix whiting with just enough water to make a thin paste. Apply the paste to buffed parts and allow the paste to dry, then wipe off the paste with a soft cotton rag. Other uses for such a paste are on auto trim, chromed parts such as bumper and wheels, glass, plastic. This procedure will remove dirt and stains.

Caution—DO NOT use shop towels, nylon, or any synthetic materials, since they will scratch a buffed finish.

Grease Stick

Grease stick, sometimes referred to as "belt grease" or "wheel grease," is a mixture of oils and greases for use as a lubricant, and, in a way, as a coolant and cleaner: it is used on abrasive belts and wheels, and, on occasion, buffing wheels. The grease is poured into paper tubes for use.

While a manufacturer or supplier may offer a number of grades of grease stick, basically speaking, there are only two kinds: (1) a light grease (tallow grease) for use when working with nonferrous metals and plastics and (2) a heavy grease for use when working with ferrous metals.

The light or "tallow" grease should be paraffin-free, because paraffin has two faults: (1) it will not flow and (2) is "gummy." Paraffin-free grease is easily cleaned. This type of grease prevents a wheel or belt from "loading" with dirt and metal, provides lubrication, and helps keep the wheel or belt from burning the part (overheating) when working on nonferrous metal. Also, on occasion, the light grease is used on a buff when buffing plastics with a compound that is too "dry" or not greasy enough.

Heavy greases, used when polishing ferrous metals, may have varying amounts of paraffin and are composed of heavier oils and greases, which are needed for the higher heat build-up when working on such metals.

A high-quality grease may cost more per pound but may be less expensive to use owing to the increased production of parts per pound of grease. This increase is due to the durability of the grease, the need for fewer applications to the wheel, and the lower cleaning costs. An additional cost factor may be the weight per tube: a 10 × 2 1/2 in. tube of the light grease will weigh approximately 1 1/2 lb; a similar tube of the heavy grease will weigh approximately 2 lb.

Buffing and Polishing Machines

Machines

The machine used in polishing and buffing is referred to as the "polishing lathe" or the "buffing lathe." These machines or "lathes" are available from as small as 1/4 hp to as powerful as 10 hp. (Heavier or more powerful machines are available; however, such machines are usually operated automatically, as part of automatic equipment.) The rpm's or shaft speed on these lathes may have a standard speed of 1800–2000 or 2200 rpm. The buffer may also be hand-held, as shown in Fig. 21.

The most common lathe has a pulley arrangement whereby the motor sets in a shell or case and a V-belt pulley to pulley ratio is established for the required rpm or shaft speed. Another type of lathe is the "motor-in-head" machine, where the shaft speed is the speed of the motor—1750–1800 or 3600 rpm.

Choosing the best lathe depends upon several factors:

1. The parts to be finished—metal or plastic; large or small.

2. The type of work to be performed—hobby work, a job shop buffing many different parts, or production runs of the part.

3. Shaft speed—constant or variable; the rpm's required.

4. Motor type—horsepower required, single or three phase; how much overload capacity is required.

The following information is offered as a guide to equipment selection.

Most shops will usually have a small 1/4 hp grinder. this grinder will often be used with a small buff. However, there is one objection to using it for buffing—the shaft is too short, which limits its maneuverability. (See Fig. 22.)

Figure 21. A typical hand-held buffer is shown with a 9 in. bonnet. Bonnets may be 7, 9, or 11 in. in diameter

Figure 22. Conventional bench grinder with a standard speed of 3600 rpm and a very short shaft, which does not allow for the versatility necessary for polishing or buffing oddly shaped parts. We suggest that it not be used for polishing or buffing or both *(Courtesy of Baldor Electric Co.)*

Small 1/4 hp machines having an extended shaft are available to handle various shaped and sized parts. (See Fig. 23.) These small machines are used mainly to finish jewelry. They are available in a single bench model, in a single-spindle or double-spindle model, or a in a complete unit comprising guards or hoods with a dust-collecting system. They have rpm's of 1725, 1800, or 3600, with 3600 rpm being the best shaft speed for any grinding, polishing, or buffing. The slower speed is better for use on low melting plastics or, in some instances, for use where a slow speed is necessary for buffing solder or when using a satin finish ("Lea" wheel) or "Scotch Brite" wheel. These machines are usually single phase, 115/230 V motors, with arbors of 1/2 or 5/8 in.

Figure 23. The Baldor motor-in-head buffer. Available from 1/4 to 10 hp, 1725 or 3450 rpm, single phase or three phase. It may be mounded on a bench or a pedestal. This machine has the extended shaft allowing for clearance for finishing of oddly shaped parts. (Note: it has NO brake.) *(Courtesy of Baldor Electric Co.)*

In finishing such plastics as the acrylics (Rohm & Haas Plexiglas or the Cyro Industries Acrylite), it is suggested that a 1 hp or smaller (1725 or 1800 rpm) buffer be used with 10 or 12 in. buffs, since these plastics burn or melt easily if too much pressure is applied; by having a low-powered machine, increased pressure will slow the machine and minimize the burning. These suggestions also apply to the buffing of telephone cases and other such plastics. (If two people are to buff plastics on the same machine, a 2 or 3 hp machine could be used with care and caution.)

In considering a heavy duty and more powerful machine, such as a 5 or 7 1/2 hp lathe, check the motor's overload capacity. Then check the size of the shaft, not the arbor; if the shaft size is under 2 3/4 in., with heavy use and using heavy wheels or buffs it will tend to whip or bend. You should consider the weight of the machine. There are inexpensive machines that weigh approximately 650 lb (using a lighter steel plate) which will give reasonably good service for several years. Other machines weigh up to

2200 lb, depending on their horsepower and their motor type, that will last a lifetime. Some examples of buffing/polishing lathes are shown in Figs. 24–26.

Figure 24. A junior or lightweight lathe useful for buffing or polishing small metal or plastic parts, where production volume is not the criterion.

Under OSHA and EPA regulations, buffing machines must be equipped with wheel guards (hoods) and suction openings for the exhaust or dust-collector system. (See Fig. 27.) The exhaust systems can be of the vacuum or cyclone type. The single-spindle, 1/4 or 1/3 hp buffer is used where the operator only uses one type of buff and one type of compound. With a double-spindle machine, two operations can be performed by the operator using different types of wheels and compounds. Some of these buffers have tapered shafts that use pin-hole center wheels, which allows the operator to change buffs or wheels instantly. Also with the taper shaft, buffs or wheels as small as 7/8 in. diameter can be used. If these small buffers have a straight shaft, taper spindles are available that will screw over the shaft, providing a bit more versatility.

There are constant or variable-speed machines as small as 1/15 hp available with flexible shafts. They have a double spindle: one end with a straight shank, the other with a tapered point. Buffs and wheels for grinding, polishing, and buffing are available from as small as 1/4 to as large as 1 in. diameter. For such small equipment, it is suggested that a jewelry supply house be contacted.

Figure 25.   The machine shown might be termed the "standard" polish-ing/buffing lathe.   Available in 5, 7, and 10 hp motors, and weighing 1000–1200 lb.

Figure 26. A split shaft and two-motor lathe. More ruggedly built, with heavier shafts, heavier bearings, and heavier body. The two motors allow for steadier production and increased versatility. It weighs 2500 lb and up.

In summary, when working on low-melting-point plastics, it is wise to use a low-power machine. Applying too much pressure while buffing will slow the machine, thus mitigating the burning or melting of the plastic.

Figure 27. (a) A guard or "hood" for abrasive belts. Made for the right or left hand, with the side hinged for quick change. (b) A guard or "hood" for the buffing or polishing wheel. The lower and upper "lips" are adjustable; the lip also has a hinged side that permits wheel changes. The upper portion is also slotted on the sides and hinged on back to allow for easier wheel changing.

In selecting a buffer, consider the motor size and also whether there will be two people using the machine simultaneously; ALWAYS have wheels of equal weight mounted on both the right and left spindles of a two-spindle machine. This conserves the bearings and also avoids bending the shaft.

While the Baldor buffer may be used on metals—both ferrous and nonferrous—it is suggested if the volume of parts and production schedule is heavy that a 5 or 7 1/2 hp heavy-duty machine be used, preferably the standard-type 1800 rpm polishing lathe, which allows for wheels up to 16 in. diameter to be used.

Semiautomatic and Automatic Machines

In considering the polishing and buffing of a large volume of parts, within a tight schedule and cost as factors, the inclination is to think in terms of automation, that is, high-speed automatic equipment. Since such equipment may be very expensive, several factors must be considered before making a decision.

* If parts are small and the contemplated production volume is large, does the part lend itself to automation or will the part need to be redesigned? What is the present production by handwork per hour, and will it be economical to automate this production?

* What types of automatic equipment are available? What is its possible production and the estimated per piece cost? The types of compounds to be used—bar or liquid? Which type of equipment to consider: a straight line machine or a rotary machine? What types of wheels and how many of them to use?

* Are the parts castings, stampings, spinnings? How many operations are involved to achieve the required finish? How much manual handling will be necessary?

* Suppose the items involved are tubing? What kind of metal? What will be the diameters? What are the lengths? How many feet of tubing per day, per week, or per month?

We have seen one occasion where the company had too many rejects and not enough good pieces to meet shipping requirements, so an automatic machine was purchased at a substantial cost. This automatic machine was very inefficient and a waste of time and money. In this case, the solution to their problem was to make changes in the hand operation after a time and motion study.

Then there was the company that had a very large automatic machine that met the required number of pieces with only one problem. There

Figure 28.

Figure 29.

Figure 30.

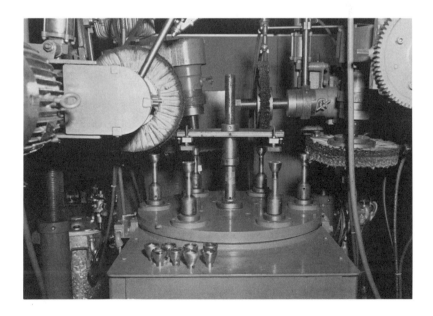

Figure 31.

was one area on the part that the machine missed, and no adjustment or change could ever reach that area, thus requiring a dozen hand polishers to finish the missed area. But the dozen hand polishers could do the complete part and meet the required production, without the automatic.

The automatic machines pictured in Figs. 28–31 (courtesy of Acme Manufacturing Co.) are examples of the endless possibilities and variations that can be built to handle any variety of parts. Volume production by automatic equipment substantially lowers the per part cost, a very important factor in today's competitive world.

Summary

1. Start slowly on any buffing or polishing operation and watch the results. Exercise care in the way you handle oddly shaped parts. Be aware that one moment of inattention can result in a serious injury!

2. Know the shaft speed or rpm of your machine. Know how to quickly calculate the surface feet per minute (sfm) your wheel is traveling; know the best sfm for the work you are doing. A wheel too fast or too slow accomplishes little.

3. Become accustomed to wearing a dust mask or respirator, which will help reduce respiratory problems.

4. Always wear safety glasses when working with abrasive belts or wheels. Your most precious possession is your EYESIGHT—protect it.

5. Do NOT wear loose or bulky clothing.

6. Work slightly below the center of the wheel (see Fig. 32). Thus, if the wheel grabs the part, the part will be thrown away from you. Also, hold the part in such a way that your hand can and will instantly release the part if it is grabbed by the wheel.

7. Do NOT use a wheel or buff with the face wider than the part to be finished. In order to then finish the part across the full surface, it is necessary to oscillate or move the part back and forth across the face of the wheel—this will make for a better and more uniform finish. As you work, you will have a tendency to work the part in an up and down motion as well as across the face of the wheel—this is good. Finally, do a pass across the face of the part—this heightens and brightens the finish.

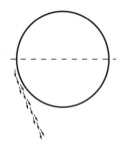

Figure 32.

Figure 33. Buff rake and handle. *(Courtesy of Manderscheid.)*

8. Oscillating the part from side to side allows any built up heat to escape. Heat build up on metal results in burns or distortions in the surface of the metal. Heat build up in high-melting-point plastics can result in burns and surface distortions and in low-melting-point plastics can cause melting and distortion of the surface.

9. Again—watch the results! Take pride in these results!

10. The long-handled platers brush is one of the handiest items to have around the shop. It may be used for brushing dirt, dust, excess compound from parts, or for scrubbing parts in the cleaning process. (This is a low cost item available from your supplier or contact the manufacturer—The Felton Brush Co.)

11. Other useful items are:
    a. The cotton or "platers" potash swab.
    b. The buff rake and handle (Fig. 33). (Caution—do NOT use a board with nails in it.)
    c. The carborundum stone $8 \times 2 \times 2$in. #24 or #36 grit.
    d. A piece of pumice (lump pumice) (available from your supplier).

Metals

Aircraft—Private

An aircraft owner wishes to remove the paint from the exterior of the plane, either to repaint or to obtain a buffed finish. The following procedures are offered as suggestions:

Contact the aircraft manufacturer for suggestions. If you are unable to follow their instructions or suffestions, consider the following items.

Some paints used on aircraft may tend to color fade, to powder, or to become very brittle or cracked. Other paints may color fade and get harder, but whatever the problem, experiment to determine the best method to use to remove the paint.

Paint remover may be suggested by the dealer or manufacturer, but such a method could be dangerous, so avoid it, if possible. Some paint strippers are illegal per state laws.

Using sandpaper by hand is a very slow and tedious process. Try a mechanical sander, using an abrasive belt, open coat #240 grit aluminum oxide. Since this powders the paint, be SURE to wear a dust mask. Using an abrasive belt with water may be tried, but will probably prove too slow and messy.

Using a wire wheel in an electric hand drill could be very rapid but could score the metal too deeply if the wire is too coarse. Also, using a fine wire may be very slow and even ineffective, since some paints or coatings become extremely hard and are very difficult to remove. CAUTION: a steel wire wheel can cause galvanic corrosion due to the dissimilarity of the metals. Check this with the dealer or manufacturer.

Once all of the old paint has been removed, the slightly rough surface will allow the new paint to adhere better.

See the following pages for alternatives to repainting the aircraft. (Note: Removing the paint from an aircraft lightens the craft by many pounds, with the result being an increase in mileage per gallon of fuel.)

The aircraft owner finds that the colors in the paint have faded and the overall finish is dull and unattractive. To restore the colors and to greatly improve th appearance, the following suggestions are offered:

1. Wash the surface with a good detergent and allow to dry thoroughly.

2. Use a hand buffer with a sheepskin bonnet, such as is used to buff an automobile. Use a compound on the bonnet specifically formulated for this purpose. Buff with a side to side motion. Do not apply too much pressure. (An up and down motion causes swirls.)

Note: To apply compound, have the buffer running, hold the bar of compound lightly against the bonnet, doing so until color shows on the buff. This will remove dirt and scatches, and at the same time, provide a highly lustrous and clean finish.

The aircraft owner wishes to remove all of the paint and have a bright finish.

1. To remove the paint, the manufacturer may offer an easily applied chemical method. (Note—some paint strippers are illegal—check it out!) Or a mechanical method using an electrical sander or an abrasive belt which results in a slightly rough finish.

2. Follow the removal of the paint, whether by chemical means or by abrasives, with a hand buffer with a shaft speed of around 2500 rpm and a sheepskin bonnet and using a fast cut tripoli for use on aluminum. This combination should remove the scratches and leave a reasonably bright finish. (Or the shaft on the right angle buffer may be long enough to hold three or four sections of 10 in. dia. ×3 in. metal center × 16 plys of 86/82 unbleached cotton muslin buffs, on which use the fast cut tripoli. By ending with a "wipe-off" motion with the buff, a reasonably bright finish is provided.)

3. Following the tripoli on the sheepskin bonnet, use a soft lambskin bonnet and a medium dry chrome rouge for the final finish. Or, using the right-angle buffer, use a wheel of three or four sections of a soft 16 ply 64 × 64 unbleached cotton muslin ×3 in. metal center 10 or 12 inch diameter with a medium dry chrome rouge. If the unbleached metal center buffs are not available, use five or six sections of 20 ply 64 × 64 unbleached loose (unsewn) cotton buffs with the medium dry chrome rouge. This should provide a highly lustrous finish.

It is important to check with the dealer or manufacturer as to the kind of aluminum used. It may be an alloy that is reasonably easy to bring to a bright finish, or it may be a pure aluminum surface that, when buffed with tripoli, may leave a dull finish. Here it becomes necessary to do some testing with other buiffing compounds, such as high-quality white diamond or other kinds of chrome rouge. Just be careful.

To clean the surface after buffing, wipe with whiting on a very soft 100or wash with a mild soap and soft cotton rags. do *NOT* use synthetic materials as they will scatch the surface.

Note: When buffing, do NOT allow the surface to become too hot, since this could adversely affect the metal. Since most aluminum now used in aircraft is clad with pure aluminum, be careful not to cut through this because doing so will leave a dull finish and corrosion will occur much faster.

Aircraft—Commercial

For the most part, commercial aircraft bodies are fabricated of aluminum, a special alloy that does not tend to oxidize. However, the aircraft does get dirty and marred in various ways.

Early aircraft were painted, and as wear occurred and time passed, they were stripped and repainted. Today, the only paint on an aircraft is the identification and company logo, and this paint is a very tough and durable plastic material. However, the aircraft body itself may be coated with a transparent material—a tough and durable plastic material upon which the line logo may have been superimposed.

The common practice of the airlines is to strip and redo the whole aircraft when the colors have faded in the logo and when a general refurbishment is required. This may include refinishing the plastic (an acrylic material), which is a labor intensive and costly operation.

Some buffing compound manufacturers make special compounds for such aircraft work. Such a compound revitalizes the colors, and removes dirt and scratches, and at the same time provides a very smooth and lustrous finish. Such a process minimizes the number of times the aircraft will need to have the finish stripped and redone.

We suggest using a 2-hp right angle hand buffer (such as is used to buff automobiles), with a 9-in. sheepskin buffing pad. While holding the buffer, running, with the one hand, lightly hold the bar of buffing compound against the pad, and as the frictional heat generated melts the compound, continue until the compound shows color on the pad, then start buffing. Recharge the pad as needed.

(Note: When buffing, oscillate the buffer from side to side, monitoring the results. Do NOT buff in an up and down motion, since this leaves streaks and skips in the finish. The oscillating action avoids building up heat in the surface of the metal, allowing it to cool. This also avoids leaving swirls in the finish.)

The buffing action provides a highly lustrous finish and, if the correct compound is used, it leaves a dry clean finish with colors brought back to their original brightness. In addition, the surface may be cleaned with whiting.

This same method of buffing may also be used on the aircraft's plastic windows. The windows may be done on both the outside and the inside without removing them. However, it may be easier to buff the windows by using a 5 or 7 in. buffing pad. As the compound is highly saponifiable, the windows may be washed using hot or warm water and a mild detergent or soap.

An alternate method of refinishing a commercial aircraft follows. If the surface is badly marred, use a right-angle hand buffer with abrasive disks, 7 or 9 in. diameter in #180 or #220 silicon carbide grain. Since aluminum will "load" the disks with metal, use a light paraffin-free tallow grease on the disk to prevent this.

By using the finer grain disks, the metal is not deeply scored or heated. To remove the scratch lines and, at the same time, provide a highly lustrous finish, use a 7 or 9 in. sheepskin pad with the special buffing compound.

If, after the buffing, it is necessary to clean dirt and debris from the surface, try wiping with a very soft (cheesecloth) cotton cloth with whiting. (We suggest that a solvent NOT be used for cleaning.)

Note: while large airlines have specially designed equipment to do the job, smaller airlines may prefer the manual method: when the steps taken and the results achieved are properly and carefully recorded, an inexpensive process can be developed that will give very acceptable results.

Aircraft Windshields

Aircraft windshields, as a matter of safety, must be optically clear—even a minute scratch or marring of its transparency can be dangerous. These scratches are caused by pebbles or rocks, birds, etc.

Aircraft windshields may be made of glass or plastic, which are specially molded to fit the contour of the cockpit. Since windshields are very expensive to replace, we suggest the following:

1. First check with the manufacturer of the plane for suggestions.

2. Do NOT remove the windshield. If the damaged area is small, as from a pebble or rock, use a right angle electric drill mounted with a fine mesh grinding wheel (for glass) (4 or 6 in. diameter × 1 in. face) and very lightly grind the scratch or fracture out, using water to keep the glass cool. Or use fine waterproof silicon carbide paper to remove the spot. (It may be necessary to use the wheel or paper in various grit sizes down to the finest finish possible.)

3. Following the grinding operation or operations, use a small diameter cotton buffing wheel with a fine mesh rouge made of an alumina. A silica or iron oxide rouge will be of no value in this operation, since they are of a lesser hardness than the glass. To maintain the optical clearness of the glass, the mesh size of the alumina rouge should be 1 micron or finer.

4. Special note: It may be possible to obtain abrasive materials of boron carbide or diamond dust—these would speed the work.

5. Glass is very hard and very hard to work on. So be prepared to work slowly and patiently—and be aware that you are working on an expensive material.

6. If the windshield is "frosted"—an abraded surface caused by sand or sleet pelting the windshield—it may be minor. Determine the severity of the condition using the following steps:
   a. Try buffing first with a smaller diameter sewed cotton buff using a medium wet chrome rouge.
   b. Try hand sanding using silicon carbide wet/dry cloth sheet in a fine mesh and if this removes the "frost"—then follow with the buffing. (Use a side to side motion.)
   c. Special note: When using disks, be sure to work from side to side and not with an up and down motion that causes swirls or rings which are almost impossible to remove.

7. Some windshields may be made of plastic—acrylic or polyurethane material. These materials are very hard, and they get harder with age. However, they may be treated in much the same way as a glass windshield.

8. To clean the windshield of grease and dirt after the buffing, use an evaporative solvent as recommended, applied with a very soft cotton cloth (cheesecloth or old used cotton sheeting); hot water and soap may also be used (for either bar or liquid buffing compound). Another cleaning method is to get some whiting and make a paste of whiting and water, wipe it on the windshield, allow it to dry and then wipe it off with a very soft cotton cloth or a chamois. (Warning—do NOT use a synthetic material, since it will scratch.)

Aluminum—to be Dyed or Anodized

When using polishing and buffing to prepare aluminum for dyeing or anodizing, the brighter the finish, the better the appearance of the finished part. (A dust mask should be used when polishing and buffing aluminum.)

In preparing the aluminum, be sure that any pits and/or scratches are removed. Depending on the condition of the metal, you may find it necessary to:

1. Use an abrasive wheel, disk, or belt, perhaps starting with an #80 or #120 grit, with silicon carbide as the abrasive. Follow each operation with an abrasive finer by about 30 or 40 points, i.e., #80 to #120 to #150 or #180 to #220 or #240. With each jump in grain size, it is best to grind slightly across the scratch lines, which avoids deepening the lines and makes the buffing easier and faster. (Caution: avoid using a hard grinding wheel, since it is so easy to have gouges or waves in the metal.) The sfm should be in the range of 5000–6000. To avoid a buildup of metal and dirt on the wheel, disk, or belt, use a light paraffin-free tallow grease as a lubricant and cleaner.

2. It is best not to use a treated buff, since the chemicals used in the treatment cause problems. Use an airway or full disk sewed buff made of 86/82 unbleached 100% cotton. For the compound, a fast cut—bar or liquid—tripoli specially formulated for use on aluminum that is to be anodized or dyed. Or use a medium dry aluminum compound made of iron-free aluminum oxide flour; consult your supplier. Use a sfm of 6000–7000. Following this buffing operation with a "wipe-off" will enhance and brighten the finish.

3. To further increase and improve the finish and brightness, you may wish to "color" buff, using a soft 64/64 unbleached cotton buff in conjunction with a fine mesh dry chrome rouge. Use a sfm of 5000–6000.

A problem that frequently occurs when anodizing or dyeing aluminum is the appearance of "lap" marks. Other problems may be "gassing" (small bubbles on the face of the aluminum) or "freckling" (brown spots). For these reasons, it is wise to be very careful in the choice of compounds, with testing of samples before purchasing any quantity being a good idea.

Many of the buffs manufactured today are made of a mix of cotton and synthetic material. When planning to buff aluminum for dyeing or anodizing, it is best to be sure that the buffs are made of 100% cotton.

While the anodizer will ordinarily clean the parts prior to anodizing or dyeing, he may insist that you furnish cleaned parts. If such is the case, you should use compounds that are easily cleaned: the liquid compound should be an emulsion easily cleaned with hot water; the bar compound should be highly saponifiable and thus also easily cleaned by hot water and soap, followed by a hot rinse. Be sure that the method of cleaning does not oxidize or dull your bright finish.

When the buffing wheel gets dirty, clean it using a buff rake lightly to remove any accumulation of dirt and compound, followed by the use of a piece of pumice stone to clean the buff and to fray the face of it.

Aluminum—Sand Castings

Aluminum foundries producing sand castings generally remove the gate, trim the flash, and clean up the casting using a band saw, a trimming die and a hard grinding wheel, or abrasive disk or belt, usually in #36 or #40 grit. Should further polishing and buffing be needed, the following steps are suggested.

1. Start with a #60 or #80 grit wheel, disk, or belt—the choice being determined by the shape of the casting. Since aluminum is a relatively soft metal, it will "load" the wheel, disk, or belt with metal and dirt. To avoid this, use a light paraffin-free tallow grease, applying it lightly to the belt, disk, or wheel as you progress. The sfm should be about 5000 or 6000.

2. To get the casting to the point where it may be buffed, use additional steps in the abrasive operation, jumping approximately 40 points in mesh size: for instance, #60 grit to #100 to #150 and so on, stopping at #180 or #220, #240, or even as fine as #320. And at each step or change in grit, slightly cross the abrasive lines: this avoids scoring the metal too deeply plus it has the added advantage of the scratch lines being more easily removed by the buffing.

3. For the buffing, use a diameter buff that will allow for a sfm of around 7000 or 8000. (Note higher speeds are not only dangerous, but usually will slide over the work, accomplishing nothing; and as the buffs wear down, the sfm lowers. Using a 16 in. diameter buff at 8000 sfm is still productive worn down to 12 in. or 6000 sfm.) For the buff, use a yellow-treated 86/82 count muslin airway with a 7 in. center plate. For the buffing compound, use a good medium dry and fast-cut tripoli. (Occasionally you will have an aluminum that is tough to cut; then try a sisal buff—laminated or treated—with a medium dry steel compound.)

4. To "finish" or "color" after buffing, use a loose or possibly 3/8 sew full disk 64/64 unbleached cotton buff. (For this operation, we suggest that you do NOT use a mixed cotton and synthetic buff.) Use, with the buff, a medium dry chrome rouge, which should provide a highly lustrous and bright finish.

5. Clean the parts using an evaporative material that does not oxidize the bright finish. Also, if possible, clean the parts immediately after buffing—this saves both time and money. Other methods of cleaning are by the use of an electrolytic process, or a hot soak soap process. Or, if these processes are not available, or the number and volume of parts does not justify their use, wipe the parts with whiting, using very soft cotton rags. (Do NOT use shop towels or a synthetic cloth, since these scratch the finish.)

Aluminum being polished or buffed is very dirty, tending to load the wheels with dirt, compound, and metal. It is suggested that when this happens, a buff rake is used lightly to remove this accumulation, which is then followed with a piece of pumice stone.

If the grease you are using contains paraffin, you will find that the paraffin, being gummy, will cause the dirt and metal to "load" quite a bit, so use both the buff rake and pumice stone to clean the wheel.

Finishing Aluminum Sheet

The various manufacturers of aluminum obtain the bauxite ore, from which the aluminum is extracted, from different areas of the world. While this aluminum chemically may have similar properties, when buffed, one lot may finish bright while another lot may finish with a dull finish.

If parts are rough and scratched and have imperfections, whatever the condition, we suggest the following procedures to finish the part:

1. It may be necessary to start with an abrasive—wheel, disk, or belt— with the least coarse grain possible. For instance, you might start with #80 or #120 grit. Your next step might be a jump in mesh size abrasive of 40 or 60 points (#80 to #120 or #120 to #150 to #180, etc.) finishing at #220, #240, or #320. Be sure to buff slightly across the lines of the preceding steps. The aluminum metal removed by the abrasive will "load" the wheel, disk, or belt; to prevent this loading, use a "tallow" grease free of paraffin as a lubricant. Should this grease contain paraffin, which is "gummy" and will not flow, you will find the "loading" will be excessive.

2. Buff with yellow-treated airway buffs, using a medium fast cut tripoli. If the aluminum is to be anodized or plated, do NOT use treated buffs; use instead a 86/82 thread count unbleached cotton buff. Aluminum is dirty and will turn buffs black; as noted previously, the buffs will tend to "load" with metal and compound. When you see black streaks on the part or the buff seems to be sliding and not cutting, the buff needs to be cleaned. To clean, first lightly use a buff rake to remove any lumps and accumulated debris, followed with a piece of pumice stone to further clean and softly fray the surface of the wheel. After buffing, a gentle wipe-off or stroke with the buffing wheel will improve the brightness of the finish. Use a sfm of 6000–8000.

3. For finishing or "coloring," use a loose 64/64 count unbleached cotton buff or a full disk 3/8 spiral sewed cotton buff and a fine mesh medium dry chrome rouge, with a sfm of 5000–6000. Clean the buff with pumice stone. Be sure to wear a dust mask during these operations. Also, since aluminum is a rapid conductor of heat, oscillate the part from side to side (and NOT up and down). By doing so, the heat is allowed to escape and a uniform finish is provided.

4. Cleaning may be done by
   a. vapor degreasing, being careful not to oxidize the bright finish
   b. an evaporative solvent (consult a supplier)
   c. hot water and soap
   d. wiping parts with a soft, frequently washed cotton rag (do NOT use shop towels or any synthetic cloth, since they will scratch the finish)

When working on round tubing, hold it lightly in your hands tilted from the horizontal, and allow it to spin (the spinning speed can be regulated or controlled by the hands) to produce a more uniform finish. When doing additional operations, at each change vary the angle of holding from the previous operation.

When working on square tubing, do NOT work in a perpendicular position, work slightly off-perpendicular, and, after each operation, work the next one at a different perpendicular position.

Barrel, Tumbling and Finishing

The barrels used in burnishing, tumbling, finishing, and plating are hexagonal (six-sided). They may be made of Bakelite, Plexiglas, hard rubber, or steel and are mounted so that they are free to rotate. These barrels may be operated dry or suspended in a liquid. Barrels range in size from as small as a quart to as large as possible.

One use of barrel tumbling is for iron castings to knock off flash, etc.; these barrels are referred to as "drums" and, more appropriately, as "rattlers." Very small electronic parts, so small that a million pieces may fit into a quart barrel, may be tumbled with a soap solution and hardwood chips to provide a finish. Several million miniature steel balls are finished by tumbling with a very fine abrasive, either dry or wet.

In addition to the barrels used for tumbling and finishing, there are the plating barrels used for specific types of electroplating. Parts too small to be finished or plated by hand, may be plated in barrels. Barrels for burnishing are made somewhat differently; they are made of cast iron and waterproof.

The advantage of using barrels, for burnishing, plating, or tumbling, is that a large volume of small parts can be worked on for very low labor and material costs.

Brass—Sand Casting

Using coarse sand for casting brass makes for a stronger and tighter (less porous), but also a rough, casting. The flash or gate on these castings may be removed with a band saw, a hard grinding wheel, or a coarse grit (#36 or #40 grit) abrasive belt. To provide a buffed finish, we suggest the following steps:

1. After removing the flash with a coarse grit wheel or abrasive belt, use successively finer grits, slightly crossing previous buff lines; starting with #80, then use #120 or #150, to #180, to #220 or #240, with a light grease on the belt or wheel. The starting grit size is determined by the roughness of the casting.

2. Buff using a medium wet tripoli on a yellow-treated airway buff. Or, as some people do, use a full disk × 1/4 sew sisal buff. If the tripoli does not have a good "cut," try a medium wet steel compound. The wheel speed should be 6000–7000 sfm.

3. To provide a highly lustrous finish, using a loose soft cotton buff with either a red jeweler's rouge or a very fine medium dry chrome rouge.

4. Clean, prior to lacquering or plating:
   a. degrease using vapor solvent or an evaporative cleaner;
   b. hot soak or electrolytic cleaner and hot rinse;
   c. for small volumes, wipe with whiting using a very soft cotton rag (old sheeting that has been laundered frequently).

Brass—Sheet and Tubing

Brass is a popular metal used for a great variety of items in the home, office, and factory. Brass is easily formed, shaped, cast, polished, plated, painted, lacquered, etc. In this connection, it may vary in quality and workability from manufacturer to manufacturer—it can be soft, hard, rough, pitted, varied in color, and easy or hard to work.

Assuming that the brass parts to be finished are rough and pitted, start by using a polishing wheel setup with #60 grit aluminum oxide or an abrasive belt. Since brass will "load," it is suggested that a paraffin-free "tallow" grease be used, which will keep the wheel or belt free of dirt and metal. A sfm of 5000–6000 should be used. In order to bring the metal to the point where it may be buffed, use consecutive steps in the abrasives, using jumps of 40 to 60 points: for instance, from #60 or #80 grit to #120 to #180 and so on until the desired finished is reached. With each step, slightly cross the scratch lines; this keeps the metal from being too deeply scored in the one direction.

When buffing, use yellow-treated airway buffs with a medium "wet" or greasy tripoli and a sfm of 6000–7500. Brass, oils, dirt, and buffing compound will load the buffing wheel, so use a buff rake lightly to remove lumps that may have formed in the wheel, and follow this by the use of a piece of pumice stone that will remove grease and dirt and fray the surface or face of the wheel. (The rake will tear the cloth, but the pumice only frays.) If black streaks appear, increase the wheel speed, slightly increase the buffing pressure, and/or try another compound.

Coloring is done to increase and improve the luster or brightness of the finish. Use a full disk soft and loose cotton buff with a red jeweler's rouge, or a very fine mesh medium dry chrome rouge with a sfm of 5000–6000. (Note: black streaks may show here also; if they do, follow the procedures mentioned previously.)

Clean the parts by vapor degreasing, hot soak cleaner, wiping with whiting on very soft cotton rags. (Do not use shop towels or a synthetic cloth, since they will scratch the finish.) Cleaning should be done as quickly as possible after buffing, since any streaks of compound and dirt left on the parts will leave white streaks in the metal that are very difficult to blend or eliminate.

Bronze Castings

Bronze is an alloy of copper and tin; nickel bronze is an alloy of copper and nickel. Regular bronze is fairly soft and easily worked, while nickel bronze is very hard and very difficult to work. Therefore, it is suggested that one know which type of bronze you are attempting to machine or polish. In either case, the following suggested procedures may be used as guidelines:

1. Foundries will do some finishing on bronze castings with #24–36 or #40 grit wheels or abrasives. Depending on how much finishing is done, you may start with #60 grit and using consecutive jumps in grit size of 40 or 60 points, slightly crossing the preceding scratch lines, going as fine as #240 or #320. Should the metal tend to load the belts or wheels, use a light paraffin-free "tallow" grease, which should keep the belt or wheels clean.

2. When buffing, if the castings are reasonably soft, use a medium wet sharp tripoli on yellow-treated cotton airway buffs and a sfm of 6000–7500. If the castings are "hard," such as nickel-bronze, try a full disk (treated or untreated) sisal buff with a medium dry steel or stainless-steel compound. (The sisal buff should probably be 1/4 in. spiral sewed.)

3. To "color" or finish the castings, use a medium dry fine chrome rouge on a loose cotton buff; a sewed cotton buff may also be used. The sfm is 5000–6000.

4. When cleaning the bronze casting, if parts are to be plated, the plater may clean the parts in a hot soak cleaner or an electrolytic cleaner, or a suitable vapor or solvent cleaner can be used. If the number of parts being processed is minimal or equipment is not available, wipe the parts with whiting, using frequently laundered cotton rags. (Do NOT use shop towels, since they will scratch the finish.)

Bronze castings have many uses—plaques, tablets, figurines, plumbing fixtures, etc. Nickel-bronze, which is durable and tough, may be used for marine items such as propellers, anchors, boat parts, some plumbing fixtures, etc. Both bronzes may be plated, polished, painted, oxidized finish, lacquered, or enameled. Be sure to use a dust mask, since the dust from grinding, polishing, or buffing can be a bit irritating.

Burnishing

Barrel burnishing or just "burnishing" is a method wherein parts are put into a hexagonal cylinder, usually made of cast iron, together with suitable media and a liquid solution. The media used may be a mixture of steel shot, fins, cones, pins, balls, etc., or such media as hardwood chips, nut shells (walnut, pecan, etc.), combined with a liquid solution–the object being to both clean and brighten the part. Do not confuse burnishing with "barrel finishing."

A common user of the burnishing process is hotels for burnishing the tableware: knives, forks, spoons, small trays, creamers and sugar bowls. The burnishing method may also be used as an inexpensive way to finish such small parts as low cost jewelry, key tags, novelties, etc., where the surface appearance is not too critical or important.

Cast Iron (Grey Iron Castings)

Iron castings, large and small, are tumbled against each other in steel tumbling barrels or drums (called "rattlers") to knock off flash and to round the edges. Often this is the only finish required.

When such castings are to be polished prior to painting or plating, you will find that they have a "skin" that must be broken or cut. (Note: Trying to polish such a casting without cutting into this "skin" is ineffective.) Therefore, we suggest the following steps:

1. Use a "hard stone" silicon carbide abrasive grinding wheel in #36 or #40 grit. (If castings are very rough, start with #24 grit.) While abrasive belts have been greatly improved, most will just "glaze" over when used on cast iron. Therefore, before investing in these belts, try some samples first. Use a sfm of 5000–6000 and no higher.

2. Once the "skin" is broken or cut, abrasive belts or "set-up" polishing wheels may be used. Following the #36 or #40 grit, go to #80 grit. Often it is good to use a lubricant on the #80 belt or wheel, and, if so, use a medium heavy petrolatum based grease. The manufacturers of such greases will use paraffin as one of the ingredients. Try samples of grease, since too much paraffin will cause your belt or wheel to "load" with metal, dirt, and grease.

3. It has become a custom, when polishing cast iron, to alternate between "dry" and "wet" wheels or belts. Since castings do vary in hardness, it is suggested that grease be tried during the various steps to develop a pattern in your processing. As you perform each step in the polishing, slightly cross the lines each time, since this avoids cutting too deeply into the metal in just one direction.

4. Proceed from #80 grit to #120 or #150, then #180 to #220 or #240. Going to #220 or #240 grit in belt or polishing wheel makes it much

easier to buff the grit line out. To do this buffing, use a sisal buff with a good steel buffing compound: fairly greasy or "wet." Here a good laminated 1/4 sew sisal buff works very well, or you may try any kind of sisal buff from an untreated full disk sisal to a four-ply 1/4 sew treated airway sisal. (Try sample wheels to determine which works best for your particular job.)

5. If castings are to be painted, the polishing may be stopped at the #120 or #150 grit step, since doing so will leave a finish that provides a better surface for the paint to adhere to. To provide a bright finish prior to plating, use a medium dry chrome rouge on a 3/8 spiral sew full disk or pieced buff using a sfm of 5000–6000. Alternately, use a loose (unsewn) full disk cotton buff with the chrome rouge, using a sfm of 6000–7000.

6. Cleaning the parts may be done with any evaporative solvent or an electrolytic cleaner. If the greases in the compounds are highly saponifiable, a hot soak cleaner can be used.

7. If none of the suggested cleaners are available or the amount of cast iron polishing is small, the parts may be wiped with whiting, using a frequently washed cotton rag. (Do NOT use shop towels or synthetic materials, since they will mar or scratch your finish.)

Deburring

Deburring is the removal of a sharp or a rough edge left from cutting or stamping a piece of metal, such as cutting tubing, trimming the edges of a casting, etc. Since many methods of deburring are available, the particular method is determined or selected based on the shape of the part, type of metal, quantity of parts, equipment available. These deburring methods utilize abrasive belts; abrasive wheels; cartridge rolls, abrasive bands, etc.; buffing wheels and compounds; barrel tumbling; vibratory finishing; router.

1. An abrasive belt may be used where the burr is easily reached or the burr is too heavy or too large to be removed by other means. Abrasive belts allow fast production.

2. Abrasive wheels are hard grinding wheels, setup wheels, Lea or satin wheels, and "Scotchbrite wheels."

3. Cartridge rolls, bands, cones are used in tight or hard to reach areas, such as inside dies or tubes.

4. Buffing wheels may be loose, sewed, or airway buffs; they are used where the burr can be easily and quickly removed.

5. Barrel tumbling can be used for heavy or large castings—aluminum, brass, bronze, iron, etc.—they are sometimes tumbled in a barrel against each other to remove flash or large burrs, or they are tumbled with an abrasive medium to both remove a burr and to round edges.

6. Vibratory finishing is a very gentle method of removing burrs or rounding edges. Since this method does not harden the surface of the parts, it is excellent for small parts prior to polishing or buffing.

7. A router is a round piece of metal varying in diameter from 3/32 to 1 1/2 in. and fluted or having the surface cut in "flutes" or "grooves,"

sometimes referred to as "round files," and partially left plain. They are used as a mandrel chucked into a hand drill to remove burrs or to remove flash from die castings. These are available from "rough cut" to very "fine cut."

8. Trimming dies.

9. A knife is used on soft metals to remove parting lines, excess metal, flash. Knives are used when other methods are not feasible owing to the shape of the parts.

Diecastings

Diecasting is a high-pressure process that forces molten metal into molds or dies; such process lends itself to manufacturing many variously shaped items.

Diecastings are usually made of alloys such as aluminum, copper and zinc, and zinc, being the most common. The castings coming from the casting machine have "flash" and, in some instances, "parting lines"—these may be removed by "trimming" dies, routers, trimming knives, abrasive belts, or wheels. Occasionally, castings may have "chill" marks, gas bubbles, or blisters—these imperfections may be major or minor. If major, the casting may be rejected, or if minor, the imperfections may be removed in the finishing process. Chill marks are caused by the die or dies not being at the proper temperature or the front part of the metal cooling too rapidly. In any event, these marks may go all the way through the casting; thus they cannot be polished or buffed out: it is cheaper to reject the part.

While most diecastings are relatively soft and easily buffed, some may require polishing by belts or wheels prior to buffing. Castings should not be tumbled prior to buffing, since this pings and hardens the surface, increasing the amount of buffing needed. To finish castings:

1. Removing the flash or parting lines or both, try to do so with a #240 or even #320 grit belt or wheel. If the belt or wheel tends to load with metal, use a "tallow" grease, free of paraffin.

2. Use a medium wet or "all-purpose" tripoli on a yellow-treated cotton airway buff—sfm 5000–7000. Since the buff will become very dirty from both the metal and compound, occasionally clean it with a buff rake and pumice stone.

3. If the parts are to be electroplated after buffing, and since the brighter the parts are prior to plating, the brighter the plating, color buff the

parts using a 100% loose cotton buff with a medium dry chrome rouge, or for brass or copper castings, one may prefer using a red jeweler's rouge on a loose buff with 5000–6000 sfm. If, in the electroplating process, there is a problem with the plate peeling or blistering, and the "color" buffing was done with a chrome rouge, change to another type of buffing compound. The peeling or blistering could be caused by the alumina in the chrome rouge.

4. Castings may be cleaned with an approved vapor cleaner, or one of following methods may be used:
   a. If the buffing compounds are water miscible, a "hot-soak" or "hot soap" cleaner may be used.
   b. Electrolytic cleaner.
   c. Wiping with whiting.

Cleaning the castings is much easier and faster if they are still warm from buffing.

Electroplating

Electroplating is a means of depositing a metal coating on another metal by an electrolytic process. The most common types of plating are:

1. Copper: For conductivity and rustproofing; often copper is "flashed" onto a part for a minimum time period to form a conductive plate and additional rustproofing prior to the next plate, which may be nickel, silver, or gold.

2. Nickel: For strength, corrosion resistance, decoration; often alloyed with another metal for both strength and hardness.

3. Chrome: For corrosion resistance, wear, decoration.

4. Hard chrome: For hardness and wear resistance of such items as crankshafts, drill bits, and dies.

5. Brass: Primarily for decoration.

6. Gold: Conductivity, decoration; used on electronic parts, electric contacts, solid-state circuits, jewelry, decorative items.

7. Cadmium: Corrosion resistance.

8. Tin: As a rustproofing on steel, as an alloy, and for decoration.

9. Rhodium: As an alloy, for corrosion resistance, and for decoration.

10. Platinum: On industrial machinery and automobiles, for corrosion resistance, and for decoration.

11. Palladium: A substitute for the more expensive platinum, for electronics, and for decoration; it is easily buffed to a highly lustrous finish.

The development of the "bright" or "brighteners" and their use over the past several years has greatly reduced the need for "color" buffing. However, the quality of these bright finishes may be greatly enhanced by a light "color" buffing, particularly of the very expensive items.

Precious Metals

There are eight metals classified as "precious," they are

1. Gold

2. Iridium

3. Osmium

4. Palladium

5. Platinum

6. Rhodium

7. Ruthenium

8. Silver

Our concern is with those most commonly used and best known in the metal polishing industry: gold, platinum, rhodium, and silver. Palladium, a member of the platinum family, is being used more frequently, owing to its lower cost.

All of these metals are easily buffed to a high degree of brilliance. They be used either as pure metal or alloyed with other metals such as cobalt, copper, or nickel.

When working (deburring, polishing, buffing) gold, platinum, or silver, we suggest some effort be made to retrieve the metal from the dust and dirt in the exhaust system, on the walls, and on the floor. If electroplating also, retrieve the metal from the sludge in the bottom of the plating tanks.

Gold

Gold is the most malleable and ductile of metals; it can be cast, rolled, beaten, laminated (gold-filled), electroplated, and has been alloyed with cobalt, copper, cadmium, nickel, and silver. Gold is used decoratively or industrially. While gold is soft and easily crafted, alloying hardens and strengthens it, also affecting its color and ease of fabrication.

Since gold, whether pure or alloyed, is fairly easy to polish and buff, we offer the following guidelines:

1. Since any use of an abrasive removes metal, to minimize this removal, depending on the condition of the metal, start with the finest mesh size abrasive possible that is consistent with good judgment. If using abrasive belts, use very soft "J" weight, and possibly no coarser than #180 or #220 grit. Or use the same grit in a "Lea" or "satin finish" wheel.

2. Assuming that the initial polishing has been done as per item 1, any abrasive lines or scratches may be removed by using a medium dry fine tripoli on a yellow-treated small airway or a multiple-layer yellow-treated cotton buff. (The assumption is that a small buffing lathe, 1/4–1 hp, 3600 rpm, is being used. In such a case, the buffs being used should be 6 in. diameter, and no more than 8 in. If the machine or lathe to be used is 1800 rpm, then 10 or 12 in. diameter buffs should be used.)

3. If the parts to be finished are in good condition and do not need an abrasive belt or wheel operation, use a high-quality fine grain white diamond bar buffing compound: such a compound will remove minor scratches and imperfections while providing a bright clean surface. use this compound on the types of buffs suggested in item 2.

4. When finishing or "color" buffing, to avoid leaving buff lines or marks on the parts, use any one of the following buffs:

   a. A very soft 60/69 thread count 100% cotton buff with one row sewing around the hole or arbor; sfm to be 5000–6000.

   b. A double nap soft cotton flannel buff.

   c. A full disk soft 100% wool buff.

   d. A soft suede material, unsewn or "loose" (which takes some experience to use well).

On whatever type of buff, use a medium dry red jeweler's rouge; use it very lightly on the buff.

  5. To clean after buffing: if the compounds used are highly saponifiable, they may be cleaned in a hot water and mild soap cleaner, electrolytically. If a solvent is to be used, consult your supplier.

Sterling Silver

Sterling silver is a soft and malleable metal used in the manufacture of table flatware, holloware, jewelry, and many other items. It is beaten, cast, hammered, molded, stamped, electroplated, etc. And because of its structure or nature, it is rather easily polished and buffed.

Owing to its softness, sterling silver parts if rough, marred, or scratched should not require the use of overly aggressive abrasives. To this end, we offer the following suggestions:

1. If using abrasive belts, start with as fine a grit as possible and a sfm of 5000. (A hard grinding or polishing wheel may cut too deeply into the metal; but, if used, use it very lightly.) Since the metal is soft, it will tend to "load" a belt or wheel; therefore, use a light tallow grease on the belt or wheel to prevent this.

2. In addition to an abrasive belt, or grinding or "set-up" wheel, a string wheel or soft felt wheel with a greaseless (Lea) compound may be used. If successive steps are necessary, be sure to buff across the abrasive lines of the previous step: this avoids scoring the metal too deeply. Use a sfm of around 5000.

3. Frequently, the parts only need to be buffed. If the metal is somewhat rough and has scratches, use a medium dry tripoli on a yellow-treated airway buff. Silver is a very good heat conductor, so it is best not to have a buff speed over 6000 sfm, and, when buffing, oscillate the part from side to side.

4. Another method of buffing the parts is by using a high-quality and fine white diamond compound on an unbleached cotton muslin airway buff or 3/8 spiral sewed full disk cotton buff.

5. To finish or "color" buff, use a full disk loose cotton buff with a good red jeweler's rouge, preferably water soluble. (Having a water-soluble rouge minimizes the cleaning of the parts.)

Silverplate

Silverplate is silver electroplated from a gray or "bright" plating solution onto a base metal—brass, copper, or steel. The "bright" plate is fairly hard and may require little or no buffing. However, to improve the quality and general appearance, it may be buffed by any of the following methods:

1. A soft double nap cotton flannel buff, loose or sewed with concentric rows of sewing 1 or 2 in. apart on which a light application of red jeweler's rouge is applied. We recommend a buff speed of 5000–6000 sfm.

2. A soft wool felt loose buff and a red jeweler's rouge.

3. A soft loose chamois buff or one with one or two rows of concentric sewing and a red jeweler's rouge, with buff speed of 5000–6000 sfm.

4. A soft unbleached cotton muslin buff (do NOT use a mixed buff of cotton and synthetic material, since it WILL scratch the finish) at a sfm of 5000–6000.

5. Sometimes the plating solution will leave a "grainy" or rough surface requiring a bit more of an abrasive action in the buffing compound. Here a high-quality white diamond compound may be used on a soft cotton muslin sewed (suggest 3/8 sew) buff followed by a loose soft cotton muslin buff with a red jeweler's rouge, again 5000–6000 sfm.

6. An "Old English," "antiqued," or "oxidized" finish may be desired. This is achieved by immersing the part in an antiquing solution (liquid sulfur solution or something similar) and then highlighting by lightly brushing with a nickel-silver wire brush under a light stream of running water. Note: When brushing under running water, use a slow rpm of 800–1000.

7. Another method of "highlighting" is by the use of a #180, #240, or #320 grit greaseless compound on a soft string buff. A loose cotton uslin buff may be used, again at a sfm of 800–1000.

8. Another type of finish is termed a "tripoli" finish. This is a dull finish achieved by buffing with a mild cutting tripoli on a soft 3/8 sewed cotton buff at a slow sfm, about 2400–3000.

Many of the buff manufacturers use a mixture of cotton and synthetic material. This kind of material may leave very fine scratches on an otherwise beautiful finish. It is suggested that discretion be used in the selection of your buffs. Check with your supplier.

Some silver plating companies will apply a "flash" coat of copper (plating enough copper to show color), then a coating of nickel, following with the silverplating. The copper plate, if "bright" copper, will not need to be buffed. However, if it does need to be buffed, a light color buffing may be done, then following the cleaning bath with a nickelplating. (The nickelplate strengthens the silver plate and improves the over all quality.) Here again, one may need or wish to lightly buff the nickelplate, then do the silverplating. Whatever steps or processing is done, when buffing the silver, take care that the buffing does NOT cut through the plate.

Guns

Refinishing a gun that has become rusty, scratched, or marred requires about the same work as preparing a gun for bluing. Guns may vary in the quality of the metal in the barrel, as the other parts may vary in the types of metal used. However, essentially the same materials and methods may be used over all the gun's parts. (Note: If there is a need to differentiate between ferrous and nonferrous metals, use a magnet: the magnet will cling to the ferrous parts and not to the nonferrous parts.)

Assuming that a single gun at a time is being finished, we offer the following suggestions:

1. Using steel wool, remove the rust, dirt, and grease from all metals parts. If there are scratches or dents in the metal that the steel wool will not remove, try a dry abrasive sheet, either aluminum oxide or silicon carbide of about #180 grit, following this with #240 grit. (If you use the abrasive grits in succeeding finer grits, buff slightly across the lines of the scratches left by the grit.)

2. If the gun has been "blued" before this cleanup, you may find that the bluing is being removed. In such a case, it may be necessary to completely remove any remaining "blue."

3. Rust pits the metal and there also may be deep scratches requiring more severe methods than the preceding operations to remove them.

4. Do not use a hard grinding wheel on a gun: it will cut too deep, leaving ripples or gouges.

5. A soft polishing wheel about 8 in. in diameter and 1 or 1 1/2 in. thick should be used. You can use a buffing wheel 1 in. thick with rows of sewing 1/4 in. apart or a medium soft wool felt wheel 6 or 8 in. in diameter and 1 in. wide. Incidentally, should you decide to purchase

a grinding machine, we suggest a 1/2 or 1 hp bench- or pedestal-mounted machine with a rpm of 1750 or 1800 with a 115/120 volt single-phase motor. Such a machine can be plugged into your house current; however, if you have three-phase, 220 volt power available, purchase a machine for this power. The machine should have a long shaft, which allows for flexibility when polishing or buffing. In addition, the rpm available will allow larger diameter wheels to be used.

6. To use a cotton buffing wheel or felt wheel 6×1 or 8×1 to remove rust and polish out some scratches, prepare the wheel by applying a greaseless abrasive compound. It is best to have several of these wheels prepared with successive grit sizes. By preparing two or three in each grit size, work does not have to stop while a wheel is being renewed. Also, by allowing these wheels to cure overnight, the surface will last much longer. Depending on the severity of the rust penetration and the scratches, you may start with #120 grit abrasive and if too fine, start with #80.

7. When working on a steel gun, you may wish to leave the #220 or #240 grit caused lines; this would be a "matte" or "satin" finish, and proceed with "bluing."

8. In addition to the regular alloyed or hard steels that we are accustomed to seeing, we now have guns made of stainless steel—a metal that cannot accept a "bluing." Since stainless steels very in hardness according to the nickel content, you will find that all can be polished with an abrasive belt or "satin" or "greaseless" wheel; occasionally, however, a stainless steel cannot be buffed. While stainless steel is resistant to oxidizing, it will discolor and stain, get dirty, and, in some instances, rust slightly. Use a medium soft felt wheel or a 3/8 concentric (single rows of sewing) sewed cotton wheel setup with #120 greaseless compound: this in all probability will both clean the gun and provide a desirable "grained" or "matte" finish. Or use a finer grain following the #120, slightly crossing the abrasive lines. (If you should wish a slightly brighter finish, apply very lightly a touch of a chrome rouge to the face of the wheel as it is running.)

9. The nonferrous parts will usually clean up by buffing with a good white diamond; if this does not work, use the "satin finish" wheel (greaseless compound) in a #120 or #150 grit, followed by buffing with a chrome rouge on a loose cotton buff. (Do NOT use a red jeweler's rouge, since it is made of a very fine mesh iron oxide it is very difficult to clean off completely.)

10. Be sure that your buffs are made of a good grade of cotton—the grades or thread counts are:
   a. 60/60—a rather loose weave; usually very soft, and excellent for color buffing.
   b. 80/92—an excellent cloth for most buffing needs; also wears better than the 60/60 or 64/64.
   c. 86/82—a tight close weave that is a hard stiff material; it is very durable, usually too hard for a coloring wheel but excellent for buffing with cutting compounds; more expensive, but worth it.
   d. Yellow-treated bufffs—these are made of the 86/82 material; the treatment makes a stiffer, faster cutting buff; it is NOT recommended for use as a coloring buff; it will hold greaseless compound.

11. Be sure that the bar compounds you use are highly saponifiable, therefore, easily cleaned. Some may be cleaned with a soap and hot water solution, while some require solvent cleaning. Greaseless compounds usually leave the work clean and free of dirt; however, if parts are to be plated or blued, they should be cleaned anyway. Detergents vary in composition and may cause streaks or stains. Also, all metals do not react the same to the cleaning agents. Check with your supplier about which cleaners to use with the compounds being supplied to you.

12. There are two items commonly used to clean a wheel. One is the buff rake, which is made of steel having sharp projecting teeth similar to a comb used to curry a horse. Since the teeth are sharp, they tear the wheel, so use the rake sparingly and only when the wheel is clogged with lumps of dirt and compound. Do not use on a grinding or hard stone wheel. Pumice stone—a volcanic lava material—will clean the grease and dirt from the grinding wheel and at the same time help restore the surface. It will also fray the surface of the buffing wheel while at the same time pulling the dirt, compound, and grease out of the buff.

13. General information on the various buffing compounds you may use follow:
   a. Greaseless or "satin" compound (sometimes referred to as "Lea" compound)—a glue and an aluminum oxide abrasive grain composition containing chemicals to keep the compound from spoiling and to dry and harden rapidly. This is available in grits from #80 to #600. (Check with your supplier as to grits and their application.)
   b. Steel compounds, sometimes referred to as "stainless compounds"—a dark gray or slightly black bar, made of a mixture

of oils and greases combined with an aluminum oxide "flour" and, possibly silica. (Be sure the compound does NOT contain any paraffin.)

c. Chrome rouge—this may be a green bar, a white bar, or a blue-tinted bar. This is made of a combination of oils, greases, and aluminas. It is easily cleaned since it is highly saponifiable.

d. Red rouge—we do NOT recommend using this compound, since it is made of iron oxide, is very soft, and very fine mesh size, making it difficult and time consuming to clean. Red rouge is of no value in buffing ferrous metal. Very fine mesh size chrome rouges are available—these cause no problem in cleaning, plating, or bluing.

e. White diamond—a very fine beige-colored silica compound, very easily cleaned, and unexcelled for buffing nonferrous metals and plastics of various kinds (e.g., fiberglass).

f. Tripoli—a reddish brown compound composed of various oils and grease, tripoli, and, sometimes, a silica or an alumina. It is used as a cutting compound on nonferrous metals and some plastics.

14. Buffing the gun butt may be done with white diamond, tripoli, or chrome rouge. However, we have found that some compounds, when used on wood, will, if wet later, show white spots that are difficult to remove. Try the proposed compound on scrap wood, sprinkle water on it, and see what happens. Some of these compounds are superb on wood and, also, on some plastics.

15. Of the many and various kinds of buffs that may be used on guns, the sisal buff (a material also known as hemp) is a very aggressive buff to use on heavy rust, to remove scratches or old bluing. Use this buff with steel compound (a dark gray or black compound). However, it should be used carefully so that it does not remove any markings, numbers, etc. For most applications use a full sisal, untreated, with no cloth covers (cloth can leave marks), approximately 10 ply, sewed 3/8 spirally. (Note: sisal buffs, if used with too much pressure against the part, can build up heat rapidly, so use light pressure. A "laminated" sisal buff is made of about six layers of cloth and five layers of sisal. Some makers will use canvas for the outside layers. Do not use these because the canvas will score the workpiece. However, if layers of muslin are used, the buff will soften on use, and can be an effective method to remove rust, debris, and old bluing more rapidly than a full cotton buff. It is always good practice to monitor a buff's performance and alter conditions as required. Finally, do not use treated sisal buff.)

16. Abrasive belts used on nonferrous metal parts, such as brass, aluminum, and diecast, will load up rapidly with dirt, metal, and grease;

to prevent this and to keep the belts clean, we suggest using a light tallow grease, free of paraffin, since paraffin does not flow, is gummy, and is very difficult to clean off completely. It can cause the bluing to be ineffective, the plating to peel, and paint or coatings of any kind to not adhere to a surface properly.

Honing

When we think of honing, we may think of the sharpening of a knife or razor blade, or possibly of the honing of the inside of a brake cylinder or piston on an automobile. Honing is an abrasive machining process widely used in the automotive and aircraft industries.

The honing process uses abrasive stones mounted in a rotating tool, which is operated, either horizontally or vertically, in a back and forth movement through a hole. This stroking movement and the rotation of the tool creates the crosshatch finish that is the characteristic of the honing process.

The abrasive honing stones used are divided into groups according to the grit size of the abrasive and the type of abrasive used for the particular metal being honed. Such abrasives as aluminum oxide, silicon carbide, cubic boron nitride, and diamond may be used for the honing. The bonding agent used to hold the grit is selected on the basis of condition of the metal and the finish desired or required. Different combinations will give different finishes in different materials.

Honing can be done on a wide range of materials: aluminum, cast iron, all types of steel and stainless steel, glass, ceramics, hard chrome plate, and such manmade materials as nylon, and in holes ranging from as small as 0.060 in. to as large as 10 ft in diameter and even larger.

Lapping

Lapping is an operation using abrasives to provide a mirror or lustrous finish; this finish is completed in such a manner that any distortion of the surface is avoided. Lapping may be performed horizontally or vertically; it may be done with a piece of wood, felt, leather, or metal with or without embedded abrasive. Ceramics, glass, ferrous and nonferrous metals and plastics can all be lapped.

Lapping abrasives may be an oil and abrasive mixture, a water and abrasive mix, or abrasive disks. Lapping compounds are available in different grit size, from a coarse grain to a very fine mesh material.

Horizontal lapping is performed with a bottom rotating plate as a base; the part to be lapped is spun in an opposing oscillating motion to ensure an even finish.

Frequently, one may have parts that require a flat surface that does not need a precision finish. In this case a round disk of metal or wood is installed on the shaft of the buffing lathe and an abrasive disk on a felt or leather wheel is then cemented or glued to it. When using an abrasive disk, either horizontally or vertically, it is best to use a grease to prevent the "loading."

Lapping compounds may be a slurry—an oil and abrasive mix—a pre-prepared material, a paste, or a bar compound. What is used is determined by the type of lapping to be done: vertical or horizontal, internal or external, type of material to be lapped, quantity of parts, and finish desired.

Marine Propellers

The marine propellers we are concerned with are those for use on smaller boats. These propellers may be made of cast bronze or cast stainless steel, either metal being very hard to polish. This being so, the following suggestions are applicable to both metals. And, assuming we have new castings, be very careful and proceed slowly.

1. Check the casting for pits and any other imperfections. Check the balance (propellers must be in balance), since this is a clue to where and how much metal needs to be removed in the polishing.

2. Because of its shape, a propeller is difficult to polish. Note that the propellers diameter is important as well as accessibility to the sides of the blades, numbering three, four, or five.

3. If using a flexible shaft for a small propeller, and to avoid making indentations in the metal, try a small diameter (3–6 in.) sewed cotton buff set up with a "Lea" greaseless Kool-kut compound, possibly #120 or #229 grit. Note: Kool-kut shows a rougher scratch line than the grit size; therefore, you might start with #320 grit. (If the imperfection or pit looks like it might go entirely through the casting, it would be best to reject it; do not take chances.) In any event, if it is necessary to perform two or three steps, be sure to buff slightly across the lines of the previous step.

4. To buff out the scratch lines and buff the entire surface of the casting, use a full disk and treated sisal buff with 1/4 in. sewing. The buff will soften from the heat generated by buffing; use a good medium dry stainless steel compound.

5. To finish and provide a very smooth lustrous surface, use a sewed cotton buff with a good chrome rouge. (A good sfm for the foregoing operations is 5000–6000.)

6. When polishing and buffing propellers on an upright polishing lathe, the main problem will be maneuvering the propeller so that the entire surface of the blades is finished:

   a. Using an abrasive belt, have the contact wheel approximately 2 in. larger than the diameter of the propeller. (As a suggestion, do not use a metal rimmed contact wheel; instead use a Diamond Crosscut contact wheel, a solid medium hard wool felt wheel, a medium soft (3/8 sew) cotton polishing wheel, a medium soft (3/8 sew) cotton polishing wheel, or sections of an airway buff—a wheel flexible enough to maneuver around and in between the blades. A hard grinding wheel will gouge the metal. Using an aluminum oxide or a silicon carbide very soft abrasive belt and, depending on the roughness of the casting, start with as fine a mesh size grain as possible. If it is necessary to use successive steps to attain a finish that is easily buffed, be sure to cross the scratch lines each preceding step. If it is necessary to use a lubricant (grease stick) to prevent a metal build-up on the belts, use a light paraffin-free tallow grease. the wheel speed should be sfm 5000.

   b. When buffing the bronze or stainless steel propeller, use a full disk plain sisal buff with a good quality highly saponifiable medium wet stainless steel compound, and 5500–7000 sfm.

   c. When coloring or "finishing," use a three or four section wheel of full disk sewed cotton buffs with the sewing determined by the diameter of buff needed, or use an unbleached cotton 86/82 count with a medium dry fine chrome rouge.

   d. When cleaning, use an evaporative solvent. Sometimes such a cleaner will oxidize a brilliant finish; therefore, an alternate method would be to wipe with whiting, using cotton cheesecloth or cotton rags that have been laundered many times.

Metal Spinning

Metal spinning is the shaping or forming of metal over a die or pattern by the process of spinning. The spinning is done on a "spinning lathe": a round piece of metal, held in place by a chuck, is shaped over a die or pattern as it spins by forcing the metal to conform to the die with a "spinning tool." This tool is a round piece of wood—oak or hickory—on the end of which a "U" shaped metal fork holds a roller or tapered point that provides the leverage necessary to force the spinning "round" to shape to the die.

As the metal spins, spinning lines appear on the part; these lines are caused by the spinning tool and their depth is determined by the amount of pressure used and the thickness of the metal being spun. The metal thickness may vary from as thin as a piece of aluminum foil to as thick as 2 in. Lubrication is usually provided by laundry soap. The removal of the spinning lines—and depending again on the thickness of the metal—may be done by machining, abrasive wheels, or abrasive belts. The finish may range from a "satin" finish to a highly buffed finish.

Nonferrous metals (brass, copper, aluminum, etc.) may be termed reasonably soft as compared to ferrous metals—thus the spinning lines may be removed by #180 grit greaseless compound, thus providing a "satin" finish, or spinning lines may be removed by buffing with a good tripoli on a yellow-treated airway buff and finished with a rouge on a loose or soft sewed cotton buff. In some instances of very heavy or thick spinnings, it may be desirable to remove the spinning lines by machining and then buffing. Should a satin or grained finish be desired, and depending on the condition of the spinning lines, it may be done with or without machining. The satin finish may be applied by a fine mesh belt, by the use of "Lea" or greaseless compound, or by the use of a "Scotchbrite" wheel.

Ferrous metals—steel, stainless steel, etc.—because of their hardness may have fairly deep spinning lines. Again, the removal of these lines and the method of removal depends on finish desired. Methods of finishing can vary from machining, use of grinding, polishing wheel, or abrasive belts, buffing, and coloring or final finishing.

The choice of steps in the use of abrasives to bring the spinnings to the point of buffing can shorten the amount of labor required to do the job. The final abrasive operation may be anywhere from #180 through #240 grit, then buff. To buff, use a good sharp steel or stainless-steel compound on a sisal buff. On the steel, you may use a medium wet or greasy compound, while stainless steel should be buffed with a dry or medium dry stainless-steel compound.

The type of sisal buff to use may be suggested or recommended by your supplier—this may be an "open face" 1/4 in. sew treated sisal, a plain (untreated) 1/4 sew sisal, or a 1/4 sew laminated sisal buff.

The final finish or "coloring" may be done by using a good quality chrome rouge lightly on a loose or sewed unbleached cotton buff. Cleaning of the dirt and grease from parts may be done in an acceptable manner—wiping with whiting, vapor solvent, soak, or electrolytic. As a suggestion, if parts can be cleaned immediately after buffing and while still warm, the results are much better.

Ferrous and Nonferrous Metal Spinning

Nonferrous metal spinnings—aluminum, brass, copper, gold, and silver—may be finished in any one of a number of ways—a bright lustrous finish, a "satin" or grained finish, an oxidized finish, highlighted or not, a painted or plated finish. To accomplish these final results, we offer the following suggestions:

1. Spinning lines can vary from very light to very heavy, depending on the thickness of the metal and the skill of the operator, the size of the part spun. If very light, they may be removed by the following.
   a. The nonferrous part should be backed by a wood block or pattern—use a treated airway buff in combination with a high-quality medium wet all-purpose tripoli. (Note: in some instances, if a very light spinning is needed, a high-quality white diamond on an airway buff or a sewed full disk 100% cotton buff.) The sfm is 6000–7500.
   b. Nonferrous metals will cause the buff to become very black with the accumulated dirt, spinning grease, and metal; when black streaks occur on the part, clean the wheel by briefly using a buff rake, following with the use of a piece of pumice stone.

c. Medium heavy spinning lines (too heavy to remove by buffing) may be removed by the use of a "lea" or greaseless wheel—using a #120 or #150 grit and then buffing. The sfm is 4000–5000.

d. The desired finish may be a "satin" or "grained" finish, as per item C, using the greaseless compound provides the "satin" finish. A finer finish might be needed, so use a #180 or #240 grit; slightly crossing the preceding buff lines and oscillate the part from side to side to produce a uniform finish. Using a light touch of a medium dry chrome rouge on the wheel will improve the appearance.

e. Heavy or coarse spinning lines as occur on thick spinnings can be removed with a polishing or "set-up" wheel or abrasive belts. The depth of the lines or the metal itself may determine the grain size to start with—this size may be as coarse as #60 grit with additional steps to reach a finish that can be buffed—this could be as fine as a #20 finish before buffing. As each operation is done, it may be advisable to use a grease or lubricant on the belt or wheel: for nonferrous metals, use a light parafin "tallow" grease; for ferrous metals, use a heavier grease. Slightly cross the preceding buff lines during each operation. The sfm is 5000–7000.

f. When buffing nonferrous spinnings, use a treated airway buff with a medium wet sharp all purpose tripoli (cleaning buff as needed). The sfm is 6000–7500.

g. When buffing ferrous spinnings, such as steel, use a medium wet steel compound on a sisal buff; this may be a full disk sewed sisal buff, a full disk treated sisal buff, a laminated (cloth and sisal) buff, or an airway sisal buff—treated or untreated. The choice of buff is determined by the quality of the metal, the shape and thickness of the part, and the quality of the finish needed as well as the production needs. The sfm is 6000–7500.

h. When buffing stainless-steel spinnings, use the same procedures as in item g with the exception that the compound should be one specially made or formulated for use on stainless steel.

i. When finishing nonferrous metal spinnings can be done with a soft cotton muslin buff, which may be loose, a sewed buff, or a soft cotton airway buff; for brass, gold, silver, and bronze spinning, a high-quality red jeweler's rouge can be used. For aluminum or nickel-silver, a fine mesh high-quality medium dry chrome rouge can be used. As the buff will rapidly become dirty, occasionally clean it with apiece of pumice stone. (Note: many people greatly dislike the use of the red rouge, so use a white chrome rouge. Viewed at arm's length, the difference in appearance is not noticeable; however, a part buffed with white chrome rouge held against one buffed with the red rouge has a noticeable difference in appearance. Therefore, it is suggested that the comparison be made before deciding which kind of compound to use.) The sfm should be around 5000–6000.

j. When finishing ferrous metal spinnings, use full disk sewed or loose cotton buffs, or, 86/82 UBM cotton airway buffs with a medium dry chrome rouge. Good steel or stainless steel compounds used for buffing will leave the parts bright, so finishing with the rouge is rapidly done greatly enhancing the final finish.

k. Cleaning both the ferrous and nonferrous parts may be done with a vapor cleaner, by an evaporative cleaner, a hot soak cleaner, hot water and soap, or by wiping with whiting powder, using very soft cotton rags. Do not use shop towels or rough cloths of any kind, since they only scratch the finish.

String buffs are ideal for providing satin finishes when using a satin or greaseless compound. However, do not even try to use one with Kook-Kut or any liquid greaseless compound.

Nickel Silver

Nickel-silver (sometimes called German silver)—an alloy of nickel, copper, and zinc—is a white, malleable and tough material. However, it lends itself to being buffed easily. Nickel-silver may be cast or rolled into sheet. It is used in many ways—as a base for plated table ware, jewelry, belt buckles, etc.

If it is used in castings such as plumbing fixtures, it may be necessary to use abrasive belts or wheels. The belt or wheel may have a tendency to "load" with metal; if so, it is suggested that a light paraffin-free "tallow" grease be used—this will prevent the "loading" and also keep the belt or wheel clean, increasing the life of the medium and increasing production. An sfm of around 6000 is excellent.

Casting of nickel-silver may be done by the centrifugal or the permanent mold process—normally such castings are relatively smooth and free of imperfections, roughness, etc., and are easily finished. Abrasive belts or wheels may be used to remove any imperfections—flash, parting lines, etc. It is suggested that aluminum oxide abrasives be used.

Buff with a medium dry all-purpose tripoli on a yellow-treated airway or a full disk sewed buff. A sfm of 6000–7000 is good. We have found that sometimes one may have a German silver item to buff and the tripoli will cause a very dark or black finish on the part. If such happens, then try a high-quality white diamond on the buff.

Color or finish the nickel-silver with a medium wet red jeweler's rouge on a 64/60 UBM loose buff. Or, an unbleached cotton airway buff of 86/82 material is excellent. If this material proves too stiff, try 64/68 UBM material. A sfm of 5000–6000 should be used.

Many people do not like to use red jeweler's rouge, because it gets on everything; an alternative is a medium dry fine mesh chrome rouge on a loose cotton buff.

Clean the parts by degreasing. If the parts are to be plated, the parts may be cleaned by the plating shop.

Nonferrous Sand Castings

Sand castings come in different shapes and conditions, with very rough surfaces to reasonably smooth surfaces.

The foundries do a certain amount of finishing, which involves the cutting off of the gate and the removal of excessive flash and metal— such finishing being done with band saws, hard grinding wheels, disks or abrasive belts, trimming dies, and/or routers. These methods leave coarse lines and a rough finish; to bring these castings to a fine finish, we suggest the following:

1. Depending on the condition of the casting, it may be necessary to start with a #36 or #40 grain size, or possibly #60 or #80. Whichever grain size is used (wheel or abrasive belt), do it as a dry operation.

2. Depending on the hardness of the casting, added steps in this polishing may be a jump of 40 or 60 points, i.e., #120, #180, then #220 or #240. As finer grains are used, the wheel, disk, or belt may load with metal; in this event, use a light paraffin-free "tallow" grease. Some castings may be mild or soft enough to buff after the #180 grit, or it may be desirable to go to a finer one, even as fine as #320. (Note: As you step from one grain size to the next, buff slightly across the scratch lines—the scoring of the metal is not as deep, consequently, your finish is better and the metal is more easily and satisfactorily buffed.)

3. When buffing, if the castings work easily indicating a good metal, use a yellow-treated 86/82 cotton airway buff with a good all-purpose tripoli. A wheel speed of 6000 to 8000 sfm is good; a higher speed can be ineffective in that the buff may "slide" instead of cutting. If this combination is not productive enough in that it is taking too long to buff, try a sisal buff, full disk, treated or untreated with either a tripoli that will work on a sisal buff or a medium wet steel compound.

4. When coloring or "finishing," which is the final buffing operation prior to a clear coating (enamel, lacquer, polyurethane, etc.), use a loose full disk cotton buff, or a soft (64/64) cotton airway buff with a suitable medium dry rouge. Use the compound lightly on the buff and a bit slower sfm speed of 5000–6000 is required.

5. When cleaning, use a suitable or a recommended solvent or wipe the parts with whiting, using a frequently laundered cotton rag. (Do not use shop towels since they will scratch the finish.)

Nonferrous metals are dirty, and will turn your buffs black; the buffs will load up with compound and dirt and leave streaks of dirt on the parts. When this occurs, clean the buff with a buff rake followed by the use of pumice stone. As you will note, the rake tears the buff, while the pumice stone will pull the dirt out of the buff and leave a softly frayed surface.

Stainless Steel

Finishes

Specific finishes have been developed for the stainless steels—they are designated as follows:

1—Unpolished Finish: A dull finish produced by HOT rolling to the specified thickness, followed by annealing and descaling.

2d—Unpolished Finish: A dull finish produced by COLD rolling to the specified thickness, followed by annealing and descaling. May also be accomplished by a final light roll pass on dull rolls.

2b—Unpolished Finish: A bright finish commonly produced in the same way as the 2d finish except that the annealed and descaled sheet receives a final light COLD roll pass on polished rolls. This is a general-purpose cold-rolled finish and is more readily polished or buffed than the #1 or #2 finish.

3—Polished Finish: An intermediate polished finish generally used where a semipolished surface is required by subsequent finishing operations following fabrication. (Note: The #1 and #2 finishes for strip (widths under 24 in.) approximate #2d and #2b sheet finishes, respectively, in the corresponding alloy types.)

4—Polished Finish: A general-purpose bright polished finish obtained with a #120 or #150 mesh abrasive, following initial grinding with coarser abrasives. (Note: In the metal polishing industry, this finish may be called either a "grained" or "satin" finish, specifying the grit size of the grain used.)

6—Polished Finish: A soft satin finish having a lower reflectivity than the #4 finish. It is produced by Tampico-brushing the #4 finish in a medium of abrasive grain and oil.

7—Polished Finish: A highly reflective finish produced by buffing a surface that has been finely ground with abrasives but the grit lines are not removed. (We would refer to this as a "satin finish.") This finish may be achieved by the use of a fine chrome rouge on an unbleached cotton muslin buff and buff lightly over the abrasive lines or by simply using the rouge in conjunction with the abrasive belt or wheel.

8–Polished Finish—This is the most reflective finish commonly produced. It is obtained by polishing with successively finer grains, then buffing extensively with a fine stainless steel buffing compound essentially to remove all the "grit" lines.

The preceding finishes have been standardized by the mills producing stainless steel and are sold on the basis of the finish. In addition, there are what are termed "proprietary finishes" developed by the mills.

Broadly speaking, the mills use two designations for their stainless steel products—"rolled" and "polished" finishes.

The mills protect the stainless steel from scratches, etc., that may be incurred in handling, shipping, and fabrication by coatings of plastic or an adhesive paper or tape, which are easily removed. Dents, scratches, and imperfections do occur however careful one may be; it is not possible to remove or repair products made from "rolled" stainless steel, but products made from "polished" stainless steel may be easily repaired or blended in.

Polishing

The condition of the stainless steel determines the operations necessary to bring the item to the desired finish. For a rough surface, possibly having some scratches, use the following procedures:

1. Starting with #60 or #80 grit; this may be a setup wheel or abrasive belt, using aluminum oxide. As you use successive grit sizes, be sure to buff across the lines—as if forming an "X"—until you either have the desired finish. This may be a #180 or #240 finish, or a very fine satin or grain finish as is obtained with a #320 grit.

2. #180 grit lines can be buffed out using a sharp stainless-steel buffing compound. However, it is easier and faster to start buffing at #220 or #240 grit lines.

3. A paraffin-free tallow grease should be used. Paraffin does not readily flow and has a tendency to "gum" up the belt or wheel. The use of grease on the abrasive belt or polishing wheel does three things: (1) it helps keep the grain from scoring the metal too deeply, (2) it provides lubrication, and (3) it prevents the belt or wheel from loading with metal. The grease may be used on each operation, but it is suggested

using it on alternate grits, starting after an initial dry operation. When grinding or polishing stainless steels, be sure to oscillate or move the part back and forth from right to left. Since heat builds up on stainless steel and does not dissipate readily, oscillating the part avoids burns.

Buffing

To buff stainless steel, we suggest the following:

1. Use an untreated 16-in.-diameter full disk sisal sewed spirally 1/4 in., with a shaft speed on the lathe of 1800 rpm—this will give a sfm of 7200. Use a medium-dry stainless-steel compound. A compound that is too "wet" or greasy will only slide on the work. It is best to use a compound specifically manufactured for use on stainless steel: it may cost a bit more, but will be less expensive in the end.

2. To color stainless steel, use an unbleached 16-in.-diameter cotton full disk buff with a dry or medium dry chrome/stainless-steel rouge. If the rouge leaves a film on the work, you might try either another rouge or a bit higher buff speed.

3. When cleaning after buffing, some solvents will tend to either oxidize your finish or leave a film. It is suggested that whiting (calcium carbonate) be used—using a cotton cloth, cotton cheesecloth, or double nap cotton flannel. (Do not use shop towels or a cloth not 100% cotton, since doing so may mar the finish.)

For more detailed information regarding stainless steels and their finishes, we refer you to your supplier (buffs and compound), to your stainless steel distributor, or write to

Committee of Stainless Steel Producers
American Iron and Steel Institute
1000 Sixteenth Street, N. W.
Washington, DC
(202) 452-7190

Buffs

Sisal buffs may be full disk, untreated, sewed 1/4 in. spirally. For a softer buff, sew 3/8 in.; for a harder buff, sew 1/8 in. Note as the buff is being used, they get hot and consequently softer.

You may choose to use "laminated" sisal buffs, usually about 11 layers of sisal with four layers of cloth between.

If the wheel becomes too full of dirt and compound, clean it by using a piece of pumice stone; this pulls the dirt/grease out of the buff and frays the surface. It is not advisable to use a buff rake, since this tears the buff too much.

For the coloring operation, a sewed or loose cotton buff may be used; this is a matter of choice or it may be determined by the shape of the parts to be buffed. Since stainless steel is very hard, airway buffs—unbleached 86/82 count cotton—may be used.

Do not use a wheel as wide or wider than the parts or metal to be buffed. Use an oscillating motion (side to side) instead of an up and down motion. Strangely enough, the oscillating movement provides an even and uniform finish, while the up and down movement leaves streaks and an uneven finish.

Vibratory Finishing

Vibratory finishing is a method of finishing using a round tub or rectangular-shaped container mounted on springs in such a way to provide both a rotary and an up-and-down motion to a mass of media and parts—such action provides a planned finish to parts and causes the parts to always maintain the same position in the media relative to each other.

Vibratory finishing is used on ferrous and nonferrous metals and plastics. The media used may be made of a ceramic material, aluminum oxide, silicon carbide, or steel—the selection of the media to be used depending on the finish desired. The media is available in many shapes and materials to provide a specific finish, be it rapid metal removal, deburring, providing a bright finish, a preplate finish, or merely improving the surface. While some media may be used dry, the most common use is with a liquid—a liquid soap, a liquid cleaner and brightener—your supplier may best help here.

There are many advantages to using vibratory finishing—a low labor cost, does not interrupt other operations, does not ping and scar surfaces, the volume finishing of small parts, and its ease of use.

Equipment for vibratory finishing ranges from as small as 3/4 ft^3 (laboratory finishing) to as large as 140 ft^3, able to handle very small parts to parts of virtually any size. And, since this method of finishing, owing to the gentle action, takes a long period (usually hours) of time, timers on the equipment are an important and necessary adjunct.

For more detailed information, contact your supplier.

Plastics

Plastics

We are dividing plastics into two categories, low melting point and high melting point, which are defined as follows:

> *Low-melting-point plastics* generally refers to those that are easily burned or melted by the frictional heat generated by buffing, approximately 300–400°F.

> *High-melting-point plastics* are those having a higher resistance to burning.

If in doubt as to which kind of plastic you will be working with, obtain a sample and experiment with it.

Begin buffing with light pressure, and slowly increase the pressure while carefully watching the results. If the part starts to burn, it will smell and ripples or distortions will start to appear; this is indicative of a low melting point. A continued increase in pressure with no visible effect on the item indicates a high melting point.

Try various buffing compounds. Some compounds are worthless on some plastics. Suppliers may recommend any number of different compounds such as tripoli, white diamond, rouge, or even steel compound. Buffs may be sewed, ventilated, sisal, or loose. In any event, the following pages on plastics contain suggestions that are the result of actual production experience and will be of value to you.

High-Melting-Point Plastics

The high-melting-point or hard plastics—thermosetting such as bowling balls, hard rubber, certain resins, Bakelite, etc.—may be brought to a high lustrous finish by buffing. Usually such plastics are hard enough that they may be polished and buffed by the same methods used to finish steel. (Note: some of these plastics require a "curing" period—as time passes, they harden and become easier to buff.)

Parts may have parting lines (from the mold), imperfections, scratches, saw lines, imperfections, or scars incurred from their fabrication, all conditions that may require the use of abrasives prior to the buffing. Depending on the condition of the parts to be finished, the following procedures are suggested:

1. The abrasive may be silicon carbide or aluminum oxide. If a wet or dry abrasive is required, use silicon carbide, possibly starting with a #60 grit using jumps in grain size of 40 to 60 points, i.e., #60–#120–#180–to possibly as fine as #320 before buffing. For each subsequent operation, buff slightly across the scratch lines.

2. You may wish to use aluminum oxide abrasives or you may wish to do "dry" operations; in this case, should the wheel or belt or disk become "loaded" with material and dirt, use a light paraffin-free "tallow" grease on the abrasive. For best results, we suggest a sfm of 5000–6000.

3. When buffing, should the abrasive operations go as far as the #320 grit, buffing may be easily done with a yellow-treated airway buff and a quality white diamond compound or use a medium dry steel compound. If the material does not respond well to the use of a white diamond or a fast-cut tripoli, then try a full disk untreated 1/4 in. sew sisal buff with a medium wet steel compound. Use a sfm of 6000–7500.

4. For a highly lustrous finish, use a medium dry chrome rouge on a loose cotton buff. Or if it is too soft, use a full disk 3/8 sew cotton buff. The compound should be used lightly on the buff, with a sfm of 5000–6000.

5. When cleaning, sometimes a "wipe-off" operation is sufficient— buffing with a clean dry soft loose cotton buff will pick up the grease left from the compound, thus providing a mild lubricating of the buff. Or use a suitable solvent as recommended by the manufacturer or supplier.

6. Wipe the parts with whiting—use 100% cotton rags from sheets, etc., that have been laundered many times. Dip the rag in whiting and wipe the parts.

7. In the event that the buffs become so loaded with dirt and compound that they become ineffective, clean the wheel by using a buff rake and a piece of pumice stone. The rake will tear lumps and chunks of dirt out of the buff and the pumice stone will pull the grease and dirt out and leave a soft frayed surface.

Low-Melting-Point Plastics

In this chapter we are concerned with the low-melting-point plastics—e.g., fiberglass, plexiglass, acrylics, injection molded, etc.. Since the low-melting-point plastics are "soft" and easily burned or "melted" when buffed, the following procedures are suggested:

1. Use soft cotton buffs, loose buffs of 60/60 or 64/64 material, with a wheel speed of 4000–5000 sfm maximum.

2. The buffing compound should be "wet" or greasy—a tripoli or a silica composition—as recommended by your supplier. Before purchasing any quantity, test the compound; it should have enough "cut" to remove minor scratches and imperfections without burning.

3. If the parts to be buffed are new and have only "scuff" marks and very minor scratches, a medium "wet" rouge may be used on the loose buff. Or, if it should be a finishing operation after the buffing with tripoli or silica compound, the rouge may be the only compound needed to provide a highly lustrous finish.

4. When cleaning, remove any dirt and grease left on the parts after buffing by the following:
   a. Parts may be wiped clean after buffing with a very soft cotton rag. (Cotton sheets that have been washed frequently.) (Caution: do not use shop towels, also synthetic materials as such will scratch.)
   b. Washing with hot water and soap or detergent. Note: if bar compounds are used, check their cleaning properties with your supplier. If liquid compounds are used, since they are water soluble, they will be cleaned easily by the use of hot water and soap.
   c. Before using an evaporation solvent, check with the plastic supplier as to the solvent to use, since some solvents will destroy the plastic. If a solvent is used, be sure to use it in the open air or in

a well-ventilated area. DO NOT SMOKE; protect the skin and eyes; use a respirator—do NOT inhale the fumes. Do NOT use near heat or an open flame.

Occasionally, it may be necessary to use abrasives (belts, wheels, abrasive paper) to remove deep scratches, parting lines, saw marks, etc.—use them with water, which prevents the plastic "loading" the medium used.

The Finishing of Low-Melting-Point Plastics

The following information will apply to all low-melting-point plastics in general and, specifically, to Acrylite and Plexiglass.

1. IMPORTANT NOTE: Under frictional heat, as occurs when grinding, polishing and buffing, a gas is given off which is toxic. As a precaution, wear a mask, preferably a respirator, that filters out both the chemical gases and dust.

2. If parts are sawed from sheet or tubing, the resultant saw lines or drill lines may be smoothed with a file, following the use of the files, it may be necessary to use abrasive paper or abrasive belts; for the best results, use silicon carbide wet or dry paper and/or belts. Unless scratches or imperfections are quite deep, start the sanding with #120 or finer grit, with successive steps going as fine as #400 or #600 grit, slightly crossing the previous lines as each step is made, which helps to minimize the time required in buffing.

3. Use a buffer with a shaft speed of 1725 or 1800 rpm. However, for a small number of parts, it is suggested using either 1/4 or 1/3 hp × 1725 rpm. If the requirement is for production runs, it is suggested that a 1 hp buffer—even with a person on each side—should be sufficient. If the buffing pressure is too high, the buffer will be slowed down, lessening the possibility of burning.

4. It is suggested using, at the approximately 1800 rpm shaft speed, a 10-in.-diameter very soft unbleached cotton loose buff with sufficient sections to provide a wheel with a face wide enough for your purpose. Or you could use a 10-in.-diameter × 72 or 80 ply buff with two rows of sewing—one row around the hole and one row halfway out to the edge in 80/92 UBM cotton. This buff provides a firmness for buffing, and if it is too hard, the middle row of sewing may be ripped out. The sfm should be between 4500 and 6000. A high-quality white diamond compound may be used to buff out abrasive lines and minor scratches and imperfections. Or a medium wet chrome rouge can be very satisfactory, providing the "cut" and high finish desired. In the

event a more lustrous finish is desired, a string buff or a double-nap cotton buff may be used with a light application of a very fine rouge.

5. On occasion, we have found that a light application of a paraffin-free tallow grease to the buff before applying the buffing compound will provide a slight increase in the lubrication and adhesiveness of the compound to the buff.

6. To clean the parts after buffing, wash with soap and hot water or with a suitable solvent; use a very soft cotton rag with whiting (a nonabrasive material); or a wipe-off operation may be done using a very soft cotton buff, such as a double-nap flannel buff.

7. Flame polishing of Acrylite and Plexiglass is often done instead of or in addition to buffing; this provides a beautiful surface. However, flame polishing requires great care to avoid leaving ripples or waves and possibly ruining an expensive piece of material. Your plastic supplier can advise you of the "how" of flame polishing and the proper materials necessary. Note: We do NOT particularly recommend oven polishing.

8. The following precautions should be taken:
   a. For low-melting-point plastics it is suggested that buffers of no more than 1 hp be used, which have a shaft speed of no more than 1800 rpm. Use light pressure when buffing.
   b. Sometimes, after you have finished your buffing, you will note what looks like lines or scratches on the finished part—these lines may be from the threads of the buff. A slight increase in the pressure as you buff should avoid this. Your buff may also be loaded with dirt and compound: if so, clean with a piece of pumice stone. The compound may also be too coarse for the material you are buffing.

Acrylite—An Acrylic

Acrylite is the tradename of an acrylic plastic material; while seemingly similar to Plexiglas, it works somewhat differently in fabricating and finishing, so care should be exercised until one is familiar with its peculiarities. Cutting, sawing, and forming should be done using means recommended by the Acrylite supplier.

If scratches and imperfections are to be removed with abrasive belts or sandpaper, use "wet or dry" sandpaper or belts, the abrasive being silicon carbide; use with water. Acrylite will "burn" rather easily, so the use of water keeps the part cool and also prevents the belt or paper from "loading" with dirt and material.

Depending on the surface condition of the Acrylite, it may be well to start with #120 grit abrasive, using successive steps in finer grits until a finish that is easily buffed is obtained: the "jumps" in abrasive grain or grit size should be about 60 points, i.e., #120 to #180 to #240 to #320 or finer, then buff.

Buffing by hand is a very laborious job; thus, a handheld electric drill or a fixed electrical buffing machine is suggested, the fixed buffing machine being the better of the two. To use the electric drill, first be sure of the rpms and the horsepower; the best sfm of the buff is 4000 and no more than 5000. Use a full disk (layer of cloth) soft 64/64 cotton muslin buff (these are available in "sections" 1/4 in. thick), in enough sections to make a wheel at least 1 in. thick. (Many electric drills do not have a shaft long enough to take any wider face buff.) A buff made of double-nap cotton flannel may also be used; these buffs should be "loose" or "unsewn."

One disadvantage to using an electric drill for buffing is that it is so easy to apply too much pressure, which can cause "burning" or ripples, or to apply too little pressure, which accomplishes nothing. (Incidentally, some buffs are made of a mixture of a synthetic and cotton; such buffs may cause problems.)

A "fixed" buffing machine, bolted to a bench or on a pedestal bolted to the floor, such as the "Baldor" buffer, which has an elongated shaft allowing the handling of variously shaped parts, and having a shaft speed of 1725 or 1800 rpms. A buffer may also be built using pillow block bearings, a shaft with pulleys, and an electric motor mounted back of or below the shaft. This machine or shaft could have two or three different diameter pulleys, which, by shifting belts, provides varying speeds.

For buffing, use a silica-based compound formulated specifically for use on acrylics, or an alumina-based rouge formulated for general finishing use on metals or plastics, which is medium "wet" or "greasy" and highly saponifiable.

As you start to buff, first fray or soften the face of the buffing wheel with a piece of pumice stone, then lightly apply compound until the compound shows on the buff, taking care to not overload the buff. As buffing is started, apply the part with light pressure against the buff, watching the results, and thus determining how much pressure to apply. (Note: Too much compound on the buff may just smear or slide, while too little can cause "burning" and distortions in the acrylic.)

Following buffing, the parts may be cleaned by washing with a mild soap or detergent and warm water, using very soft and frequently washed cotton rags. (Rags containing any synthetic material will cause scratches.) Another way to clean parts is by using "whiting" powder on very soft cotton rags. It is best to avoid solvents for cleaning.

Two notes of caution should be observed: (1) always use face masks when working with acrylics and (2) do NOT try to flame polish, since Acrylite is very combustible.

Plexiglas—An Acrylic

Plexiglas is thought of as being a hard material and may be treated as such: this understanding about it can be an expensive mistake.

In fabrication, Plexiglas sawed or cut will have a rough surface and edge that have to be removed. As this material is what may be termed "soft," the following steps may be used to bring the parts to a beautiful finish:

1. To remove rough saw marks and imperfections, use a "wet or dry" silicon carbide abrasive cloth or paper, or silicon carbide "wet or dry" abrasive belts with water, being careful NOT to generate too much heat. Starting with a #60 grit, use successive steps, i.e., #120 to #180 to #240, or even finer, slightly crossing the scratch lines with each step. The sfm should be around 4000.

2. Following the last abrasive step, buff with a soft 80/92 unbleached cotton muslin buff at a sfm of approximately 4000; the buff should be loose or have just one or two rows of sewing in addition to the sewing around the hole. The buffer should have a shaft speed of 1725 or 1800 rpm. A 10-in.-diameter buff on a shaft of 1800 rpm = 4500 sfm—an excellent speed providing the buffing compound used has sufficient lubrication. Another type of buff is an airway buff with a 3 in. metal center made of a soft 80/92 unbleached cotton muslin; it is excellent providing due attention is paid to the results or finish. Here a white-diamond-type compound is good if the mesh size of the abrasive is not too rough or coarse, or a medium "wet" chrome rouge is used. However, since compounds do vary from manufacturer to manufacturer, it is wise to try compounds for quality of finish before buying. Note, again, buff cloth containing partial synthetic and part cotton can cause scratches on the finish, so be very careful.

3. Cleaning parts after buffing can also be a problem. The supplier of the Plexiglas will supply or recommend a cleaner. Should the buffing compound used be a highly saponifiable liquid compound or bar compound, the parts may be washed using a light soap and hot water. If rags are used to do the washing, be sure that they do not contain any synthetic material.

If flame polishing is used, be sure to check with your supplier: we have found that various fabricators recommend different mixtures of gas and oxygen. Flame polishing does save a lot of time, but care must be exercised to avoid ripples and other imperfections. However, flame polishing does leave a beautifully clear, clean finish.

Cultured Marble

Cultured or synthetic marble is a mixture of resins and selected powders or fillers. This mixture is formed into vanities, mantels, various kinds of tubs, showers, and various decorative items for the homes, offices, buildings, etc. These items, when released from the mold, may have minor imperfections, scratches from the handling, and "orange peel" skin.

Some imperfections may be minor enough to be removed by a sanding operation or by buffing. To remove these and to enhance the beauty of the finished product, we suggest the following procedures:

1. To remove imperfections use a wet sandpaper (silicon carbide), possibly #150 or #180 grit, sanding by hand, and being careful to restrict the operation to as small an area as possible. (Note: the scratches from these grit sizes will be removed by the buffing; minimize sanding as much as possible.)

2. To buff—use a right angle hand buffer—air or electric operated and having a 7 or 9 inch diameter rubber disk over which a sheepskin bonnet may be mounted. See Fig. 34. (Some sheepskin pads have a heavy canvas backing and are made having an arbor threaded to screw on to the right angle buffer.) Have the buffer running full speed, hold the compound bar lightly to the buff—as the frictional heat thus generated will soften the compound and will adhere to the bonnet or pad—once the compound thus shows, start buffing. As you buff, oscillate it from side to side, watching the results as you work. Do not use an up and down motion or work in a circle, since this makes for swirls.

3. If using a liquid compound or paste, using a swab, apply to the part, and then buff. The buff will pick up the compound. The finished or buffed part should have a silky smooth and lustrous finish. If any cleaning is needed or necessary, wipe with whiting powder using very

Figure 34. Buffs for cultured marble. The buffing pads illustrated are available in 7 and 9 in. diameters. These have a hard canvass backing and are made with a center that is screwed onto the shaft of the hand buffer. Lower left hand buff is a 50/50 natural and synthetic material: this is to be used for a fast cut operation—removal of sand paper scratches or imperfections or both. Because it will leave marks on the product, it should not be used as a finish buff. Lower right hand buff is made of 100% twisted wool. This buff is used for a mild cutting action of minor imperfections and fine sandpaper scratches. This pad, too, should not be used for a final finish, since it leaves the finish dull. Top center buff is the softer pad: made of a 100% twisted wool and provides a highly lustrous and silky smooth finish *(Courtesy of B & H Industries, Ltd., Mason, MI.)*

soft rags or cheesecloth. Do not use a coarse material, such as shop towels, because they will scratch the finish.

4. Buffs will get very dirty from the compound and marble—when the buffing is leaving streaks, these may be cleaned by using a piece of pumice stone. However, once the buffs are too dirty to use, they may be washed. (Check with your buff supplier.)

Fibreglass

Fibreglass is made of glass filaments woven into a cloth. Layers of this cloth are laminated into various thicknesses and bound by resins to form such items as molds, boats, skis, surfboards, various kinds of tubs, and so on, with new uses being constantly found. Such laminations have to be cured. After these items are cured, they may have a rough surface, minor imperfections, and/or scratches requiring buffing to obtain an acceptably smooth and lustrous surface.

The resins bonding these laminations are easily "burned" or melted by the frictional heat generated by the buffing operation. In order to avoid this "burning" or melting, great care should be exercised in the selection of a buffing compound, which should be one especially formulated for use on fibreglass. If the compound is too "wet" or "greasy," it will slide instead of cutting, and, also, if too "dry," it will "burn" and not cut. In addition, the compound should be highly saponifiable or easily cleaned by a suitable evaporative solvent or by washing with a soap and hot water.

If the items are to be buffed on a standard upright or vertical buffing lathe, we suggest using a soft cotton full disk buff, 60/64 or 64/68 thread count, loose (no sewing other than around the arbor hole or slightly firmer with two rows of sewing, or, if an airway or ventilated buff, of the 60/64 or 64/68 thread count, using a rpm speed of 4500 to 5000). If items are to be buffed horizontally with a hand-held right angle buffer, using a sheepskin bonnets or pads, the following procedures should be followed:

1. Have buffer running full speed, apply compound lightly to the wheel until "color" shows, no more (too much compound is ineffective and just a waste). (Note: The frictional heat generated as bar compound is held against the wheel will melt it and the binders in the grease will cause it to hold to the buff. See Fig. 35.) Watch the results and apply additional compound as needed. If the compound is a paste or

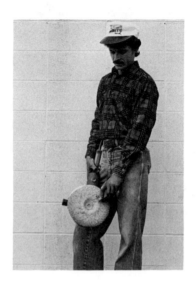

Figure 35.

       liquid, it may be applied or sprayed on to the buff by air, or applied
directly to the part by a swab.

2. Following buffing, clean the parts with an evaporative solvent or by
washing with a soap and hot water.

A mold release compound may have been applied that does not need
to be removed. The vertical buffs or buffing wheels get dirty from both
the compound and dirt; the wheel may be cleaned with a piece of pumice
stone, followed by a light use of the buff rake, or a rake, pumice stone,
rake sequence. When cleaning a bonnet or pad, do not use a buff rake;
instead, use a piece of pumice stone.

Once a bonnet has become just too dirty to continue using, wash it:
liquid compound washes out easily, while a bar compound may require a
strong soap and hot water solution. (The bar or liquid compound should
be formulated specifically for use on fibreglass.)

If refinishing items made of fibreglass, which are both very dirty and
oxidized (painted): to clean and restore the colors, use a "fibreglass"
rouge such as the "WP" rouges manufactured by The Sattex Corporation,
Vernon, CA.

Index

abrasives 23
  belts, 28
  bonded, 14
  cartridge rolls, 28
  diamond, 25
  disks, 29
  flours, 26
  grains, 24
  wheels, 29

Acrylite, 141

aircraft,
  commercial, 76
  private, 73
  windshield, 78

aluminum
  anodized, dyed, 80
  castings, 87
  sheet, tubing, 84

barrel
  burnishing, 28
  finishing, 86

bonnets, 144

boron carbide, 26

brass
  castings, 87
  sheet, tubing, 88

bronze
  castings, 90
  sheet, tubing, 90

brushes
  platers, 68
  wire, 19

burnishing, method, 92

buffer
  automatic, 66,67
  Baldor, 61
  bench, 60
  hand-held, 60, 146
  upright, 63

buffing, summary, 69

buffs
  airway, 5
  bobs, 10, 11
  finger, 14
  fluted, 10, 11
  goblet, 10, 11
  introduction, 13
  jewelers, 7, 8
  knife edge, 6
  loose, 8, 9
  pads, 5
  pocket, 11
  rake, 70
  razor edge, 6
  sisal, 9

castings
  aluminum, 82
  brass, 87
  bronze, 90

die, 97
iron, 93
steel, stainless steel, 127

cement, polishing wheel, 36

cleaning, 54, 55

compounds, description
   crocus, 44
   emery cake, 44
   fiberglass, 45
   liquid, 46
   marble, 45
   rouge, 44
   atin, 45
   steel, stainless steel, 43
   tripoli, 42
   white diamond, 43

deburring, 95

diamond, 25

diecastings, 97

dies, trimming, 123

electroplating, 99

emery, 25

emery cake, 44

fibreglass, 145

flours, 26

glue
   kinds, 34
   pot, 34

gold, 102

grease, 59

greaseless, 38

grinder, bench, 61

guns, 108

guns, spray, 46

honing, 113

lambskin, 144

lapping, 114

lathe
   buffing, 68
   automatic, 66
   semiautomatic, 66
   summary, 69

liquid compounds 46
   schematic, 47
   trouble shouting, 51

marble
   cultured, 143
   synthetic, 143

marine, 115

metals
   alloys, xv
   pure, xv
   precious, 101

nickle, nickel silver, 121

plastics,
   Acrylite, 139
   finishing, 137
   high-melting point, 134
   low-melting point, 136
   Plexiglas, 141

plating, 99

points, 14

pumice
   lump, stone, 27
   powder, 27

polishing wheels, 12, 29

rake, buff, 71

rouges, 44
   chrome, 44
   jewlers, 44

sheepskin, 144
   sand castings, 123
   satin compound, 38
   schematic, 47

shop
  in-plant, 3
  job, 3
  polishing, 3

silicon carbide, 25

silver
  german, 121
  plate, 106
  sterling, 104

spinner, work-holding, 21

spinning, metal, 117

stainless steel, 125
  buffing, 127
  finishes, 125
  polishing, 126

tallow, 58

tripoli, compounds, 42

tumbling, 86

vibratory, 129

wheels
  abrasive, 29
  contact, 30, 31, 32
  grinding, 16
  polishing, 12, 29
  Scotch-brite, 17
  set-up, 16
  wire, 16

windshield, aircraft, 78

white diamond, 43

zirconium, 26